OXFORD STUDIES IN PHYSICS

NEUTRON STARS

BY

J. M. IRVINE
Reader in Theoretical Physics,
University of Manchester, England.

1478
1978

CLARENDON PRESS · OXFORD

Oxford University Press, Walton Street, Oxford OX2 6DP

OXFORD LONDON GLASGOW
NEW YORK TORONTO MELBOURNE WELLINGTON
IBADAN NAIROBI DAR ES SALAAM LUSAKA CAPE TOWN
KUALA LUMPUR SINGAPORE JAKARTA HONG KONG TOKYO
DELHI BOMBAY CALCUTTA MADRAS KARACHI

ISBN 0 19 851460 3

Printed in Great Britain by
Thomson Litho Ltd, East Kilbride, Scotland.

CONTENTS

PREFACE

THE concept of a neutron star as one of the possible end-points of stellar evolution was introduced by Landau, so the story goes, on the day that the news of the discovery of the neutron reached Copenhagen from Cambridge. By the end of 1933 the idea that neutron stars might be produced in the cores of supernovae had been advanced by Baade and Zwicky; and six years later the work of Oppenheimer, Volkoff, and Tolman had led to the first estimates of the maximum mass of a neutron star.

For the next thirty years neutron stars remained a curiosity outside the mainstream of astrophysics and astronomy. The problem was that neutron stars were predicted to be dimensionally small, having radii in the range 10 to 100 km. All the estimates indicated that they should cool extremely rapidly, reaching a surface temperature of less than 10^6K within a few years of their formation. Thus the luminosity of neutron stars was predicted to be so low as to render them undetectable.

Matter accreting on to the surface of a neutron star would be greatly compressed, and consequently its temperature would rise sharply, resulting in the emission of x-rays. There was brief speculation that the observation of compact x-ray sources in the 1960s heralded the discovery of neutron stars; but detailed calculations indicated that these sources were not sufficiently compact and were more probably white dwarfs.

Towards the end of 1967 a Cambridge radio astronomy group working under the direction of Anthony Hewish discovered the first pulsar. The short period between the pulses and the much shorter duration of each pulse argues for an extremely compact source. Almost immediately Tommy Gold suggested that pulsars could be rapidly rotating, magnetic neutron stars. The subsequent discovery that two of the most rapidly pulsing neutron stars were associated with remnants of the supernovae which gave rise to the Crab and Vela nebulae seems to confirm Baade and Zwicky's prediction.

The discovery of pulsars and their interpretation as neutron stars coincided with a period of advance in our understanding of gravitational collapse and in the development of many-body theory. The combination of these events has led to an explosion of interest in neutron stars over the past decade. Studies of neutron star models have revealed an astonishing richness of structure, much more complicated than the simple neutron fluid originally envisaged by Landau.

This is a book strictly about neutron stars and particularly their structure. It is not a book about pulsars. Where it seems appropriate I have called on information obtained from the observation of pulsars to motivate the discussion or justify certain features of the neutron star models. I have not discussed in any detail the dynamic formation of neutron stars. This is most likely an explosive sequence of events and is probably associated with supernovae, although there may also be quasi-static routes to the neutron star state. However, at the present time there is no detailed understanding of how neutron stars are formed. I have also not discussed in any depth how pulsars pulse. Here again there is no consensus view on the subject, although there exists a plethora of models. It is not clear whether the pulses originate at the surface of the star, or much further out in the magnetosphere, or indeed whether the same mechanism is responsible for the pulses in all pulsars. Finally, I have not discussed the search for neutron star x-ray sources and binary pulsars.

We should bear in mind that pulsars may not be neutron stars. If this is the case it will simply serve to remind us of the wonder of nature. Certainly, to date, man has not had the wit to invent an alternative explanation.

My own interests in neutron stars have grown slowly over a number of years, but were greatly heightened, appropriately enough in Copenhagen, by a meeting to discuss relativistic astrophysics organized by Canuto in the spring of 1974. They developed in the pleasant atmosphere provided by the nuclear theory group at the University of Maryland during the autumn of 1974 and the stimulating surroundings at SUNY Stony Brook in the spring of 1975. The final version of the typescript was completed while I was visiting Michigan State University in the spring of 1977.

I have received considerable enlightenment from Gordon Baym,

Hans Bethe, Gerry Brown, Leonardo Castillejo, Mal Ruderman, and Phil Siemens. My colleagues in Manchester have always provided a useful sounding board, and I would like to acknowledge the assistance I have had from Ray Bishop, John Evans, John Owen, Ian Rodgers, and Mike Strayer.

I am particularly grateful to John Evans, Bruce Winterbon Dennis Sciama, and John Miller for their critical reading of the manuscript, which leaves me totally responsible for the remaining errors and omissions. Heather Kimber and Abbey Rosemont did an excellent job in preparing the typescript, and the staff of the Oxford University Press with their usual efficiency lightened my load considerably.

This book is dedicated to Ritchie, for whom the stars hold great wonder.

Manchester J.M.I.
May 1977

1

INTRODUCTION

1.1. Resumé of stellar evolution

THE only completely universal interaction known in physics is gravity. The story of stable or quasistable stellar objects is the story of the competition between other physical effects and the universal pull of gravity. Without this competition no stable stellar objects other than black holes could exist.

Opposition to gravity may be provided by kinetic effects or by the other fundamental interactions of physics, i.e. electromagnetic and nuclear forces. Angular momentum is a kinetic effect, and the resulting centrifugal barrier can prevent gravitational collapse and provides a measure of stability in, for example, the solar system or the galaxy. Other examples of kinetic processes opposing gravity are provided by the normal hydrodynamic pressure and radiation pressure in a main sequence star and neutrino pressure in white dwarf or neutron stars. For the subject matter of this book a particularly important kinetic effect is provided by the fermion nature of matter. Fermions obey the Pauli exclusion principle, and thus no two fermions can be in the same state. This implies that, even at absolute zero temperature, there is a distribution of non-zero momentum states occuped by the fermions and hence a resulting pressure, i.e. the Fermi pressure or degeneracy pressure, which will oppose the pull of gravity.

Most kinetic effects lead only to quasistability. For example, angular momentum can be lost through the action of normal hydrodynamic viscous forces, or be radiated away gravitationally, or in the case of magnetic systems it may be lost into electromagnetic radiation. Hot systems will in general radiate away their free energy in the form of photons or neutrinos. Eventually the sources of free energy will become exhausted.

How the temperature of a stellar object responds to this struggle against gravitational collapse depends on the equation of state for the material of which it is composed. As the object collapses the gravitational forces become stronger, and if the collapse is not to become catastrophic the pressure gradients must rise at least as rapidly. In the case of a main sequence star, as the pressure rises, so does the temperature. However, as the star approaches stable equilibrium the pressure becomes essentially independent of the temperature and the star will eventually cool down.

The sequence of collapses is a dynamical process. The collapse may be extremely violent, resulting in processes in which much of the original mass of the star may be blown off. Thus, if a main sequence star is too large, the radiation pressure may rise too rapidly, leading to an instability, or infalling matter may bounce off the degenerate core of a star leading to shock waves with an associated sharp rise in density and temperature, which may result in thermonuclear explosions, or degenerate stars may be so massive that neutrino pressure may drive off the outer layers of the star. The stable, cold objects that can be formed are then the embers of such explosions, and their mass limits are as much prescribed by the dynamics of their formation as by the statics of the resulting cold, stable structure.

In the case of meteorites the mass is so small that gravitational effects are insignificant and their structure is dictated by the ordinary forces of solid state physics. For a gravitationally self-bound object, if these same forces are sufficient to halt the gravitational collapse, the end-point is a planet. If the mass is too great, a planet will collapse under its own weight; but this process may be halted by the electron degeneracy pressure, in which case the end-point is a black dwarf star, i.e. a cold white dwarf. If the mass is even greater, the electron Fermi pressure may not be sufficient and the collapse may continue until halted by the strong nuclear forces or the baryonic[1] Fermi pressure; in this case the end-point is a neutron star. If the mass is too great for a stable neutron star to form, then the collapse to a black hole is predicted.

---

[1]The term hadron is used to describe all particles which experience the strong nuclear interactions. Baryons are the fermion hadrons as distinct from mesons which are the bose hadrons.

We stress again that the resultant stable stellar objects are the products of the dynamics of the collapse; and hence we are not implying that all black holes are more massive than all neutron stars, which in turn are more massive than all black dwarfs, etc. For example, the collapse to form a neutron star may be so explosive that much of the matter of the star may be blown off and the resulting core, which becomes the neutron star may be only a small fraction of the initial mass of the star.

In this book we shall not treat in any detail the dynamic formation of neutron stars, but will concentrate on the possible properties of neutron stars once they have been formed.

In elementary physics we learn that a gas will expand to fill the volume available to it; and while this is true in the laboratory it is not true on a cosmic scale. A gas cloud becomes unstable against contraction if the gravitational binding energy exceeds the thermal energy of the gas. For a mass M of a perfect gas occupying a sphere of radius R at temperature T the system will be unstable against gravitational collapse for

$$\frac{GM^2}{R} \gtrsim NkT \tag{1.1}$$

or

$$R > \frac{V_s}{\sqrt{\rho G}} \tag{1.2}$$

where V_s is the velocity of sound

$$V_s = \sqrt{(P_G/\rho)} \tag{1.3}$$

for a gas of uniform density ρ and pressure P_G.

The inequality (1.2) presents us with a natural minimum time period τ for any density oscillation of the system

$$\tau = R/V_s > 1/\sqrt{(\rho G)} \tag{1.4}$$

Note that this is also the period of a satellite on a minimal orbit and as such is also a characteristic period for rotational features of the system. In Table 1.1 we present some characteristic periods and densities for various stellar objects.

TABLE 1.1

τ	$\rho\,[g/cm^{-3}]$	
~ hour	10	planet, main sequence star
~ second	10^7	white dwarf star
~ millisecond	10^{13}	neutron star

The condition for stellar equilibrium is (Fig. 1.1)

$$\frac{dP}{dr} = -\frac{GM(r)\rho(r)}{r^2} \tag{1.5}$$

in Newtonian gravitation theory, where $M(r)$ is the mass inside the radius r. In order to obtain various order of magnitude estimates we shall frequently assume a uniform density ρ and approximate $\frac{dP}{dr}$ by $\frac{-P}{R}$.

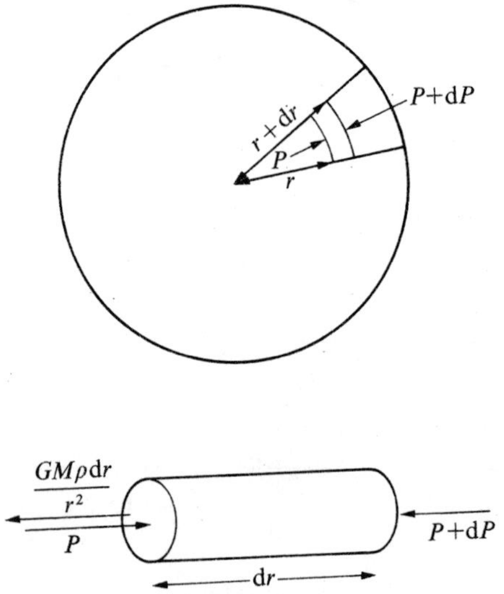

Fig. 1.1. A small element of stellar material of unit cross-section in which the pressure gradient balances the gravitational forces.

For any system the Helmholtz free energy F is given in terms of the internal energy E, the entropy S, and the temperature T by

$$F = E - ST \tag{1.6}$$

whence the pressure P is defined by

$$P = -(\partial F/\partial V)_T. \tag{1.7}$$

Hence for a cold object the pressure is given by the negative of the internal energy density. In a highly collapsed situation, such as a black hole or a neutron star, the contribution of the internal energy density to the gravitational potential must also be taken into account.

There is, of course, a limiting value for the velocity of sound since it may not exceed the velocity of light c. Indeed, to have a positive definite stress tensor the limit is $c/\surd 3$. Thus, from equation (1.2) we deduce a critical radius R_{crit}.

$$R_{\text{crit}} = c/\surd(\rho G) \simeq MG/4c^2. \tag{1.8}$$

This should be compared with the event horizon, Schwarzschild radius, or black hole radius for the mass M

$$\mathcal{R} = 2MG/c^2. \tag{1.9}$$

If a mass M collapses to an object within a radius less than $\mathcal{R}$ it will continue to collapse forever, and no force in the universe can halt it.[1] General relativistic effects will be of the order

$$\delta \simeq \mathcal{R}/R. \tag{1.10}$$

Starting with a primitive cold gas cloud, we see that it

---

[1]Very recently there has been speculation about quantum field effects in gravitational theory, and it has been suggested that black holes may radiate particles and indeed may do so explosively. Such considerations may in future negate the assertion made above.

will start to collapse under its own gravitational field, if it is large enough. The gas will then be compressed and its temperature will rise until the inequality (1.1) is violated. As the temperature rises the gas will begin to radiate and the radiation pressure will supplement the normal hydrodynamic gas pressure in opposing the gravitational collapse. However, if the gas cloud is too large, the radiation pressure may rise too rapidly (remember it increases as the fourth power of the temperature) causing an instability.

An estimate of the mass of a star in the quasistatic main sequence collapse mode can be obtained by requiring that the radiation pressure P_R

$$P_R \simeq kT\left(\frac{kT}{\hbar c}\right)^3 \tag{1.11}$$

should not dominate the gas pressure

$$P_G \simeq \left(\frac{\rho}{\mu}\right)kT \tag{1.12}$$

where μ is the mass of a gas molecule. Assuming that the radiation and gas pressures are comparable and using (1.11) we have

$$M \lesssim 2/3\left(\frac{\hbar c}{G\mu^2}\right)^{3/2}\mu. \tag{1.13}$$

If we assume that the star is composed of monatomic hydrogen we have

$$M_c = \left(\frac{\hbar c}{G\mu^2}\right)^{3/2}\mu \simeq 3 \times 10^{33} \text{ gm} \simeq 3/2\ M_o \tag{1.14}$$

where M_c is known as the Chandrasekhar mass (see eqn (1.28) below) and will serve as a useful unit of mass in our discussions and M_o is the mass of the sun.

Assuming a uniform density $\rho = 1$ g/cm^{-3}, so that

$$M \simeq \rho R^3 \tag{1.15}$$

the mass in (1.14) gives us a scale size for main sequence stars of

$$R \simeq 10^{11} \text{ cm.} \tag{1.16}$$

If the temperature of the collapsing gas cloud is high enough, thermonuclear reactions can take place, and such fusion processes are the source of energy in main sequence stars. Since the thermonuclear reactions are induced by the strong nuclear interactions which have a range of $\sim 10^{-12}$ cm at which distance the Coulomb energy of two protons is ~ 0.1 MeV, we may expect thermonuclear reactions to set in at a temperature of a few keV per particle, corresponding to $\sim 10^7$ K. Reactions involving more highly charged nuclei require higher temperatures in order to overcome the increasing Coulomb repulsion.

The first and longest stage in the life of a main sequence star is the burning of hydrogen to form helium which occurs at $\sim 10^7$ K and densities ~ 10 g cm^{-3}. The exhaustion of hydrogen will not in general occur uniformly throughout the star, but will first be achieved in the hotter, denser core of the star. What happens next depends on the mass of the star. For low mass stars the temperature in the core may not be sufficient to ignite helium burning. The core collapses becoming degenerate, i.e. supported by the electron Fermi pressure, and the star moves into the white dwarf stage.

In more massive stars the release of further energy by continued nucleosynthesis may take place. With a star containing a mixture of hydrogen and helium the most obvious nuclear fusion products would appear to be ^{5}Li and ^{8}Be, but neither of these nuclei are stable. However, ^{8}Be is almost stable (half-life 10^{-16}) and a finite concentration of ^{8}Be, resulting from a dynamic balance in the capture and dissociation reactions, may be established. The ^{8}Be can capture another helium nucleus to form an excited state of ^{12}C which then decays to the ground state by gamma emission. This three-body resonance in ^{4}He leading to the formation of ^{12}C is extremely temperature dependent and occurs at $\sim 10^8$ K.

Having formed ^{12}C the door is open to a whole series of helium burning processes

$$^{12}\text{C} + {}^4\text{He} \rightarrow {}^{16}\text{O} + \gamma$$

$$^{16}O + {}^{4}He \rightarrow {}^{20}Ne + \gamma$$

$$^{20}Ne + {}^{4}He \rightarrow {}^{24}Mg + \gamma$$

etc., with successive reactions begin triggered as the temperature rises sufficiently to overcome the Coulomb barrier. At higher temperatures still we have ^{12}C burning, principally

$$^{12}C + {}^{12}C \rightarrow {}^{20}Ne + {}^{4}He$$

$$\rightarrow {}^{23}Na + H,$$

(at $\sim 8 \times 10^{8}$ K); followed at $\sim 1.5 \times 10^{9}$ K by the endothermic photo-disintegration of ^{20}Ne

$$^{20}Ne + \gamma \rightarrow {}^{16}O + {}^{4}He$$

which acts as a source of free alpha particles for the exothermic ^{4}He capture reactions

$$^{20}Ne + {}^{4}He \rightarrow {}^{24}Mg + \gamma$$

and

$$^{24}Mg + {}^{4}He \rightarrow {}^{28}Si + \gamma$$

This neon-burning chain is exothermic giving a net positive energy output; next at $\sim 2.0 \times 10^{9}$ K we have ^{16}O burning, principally

$$^{16}O + {}^{16}O \rightarrow {}^{28}Si + {}^{4}He$$

$$\rightarrow {}^{31}P + H$$

$$\rightarrow {}^{31}Si + n$$

etc., leading to the formation of the heavy elements. If the star is massive enough an onion-like shell structure may develop with different nuclear synthesis processes occurring in

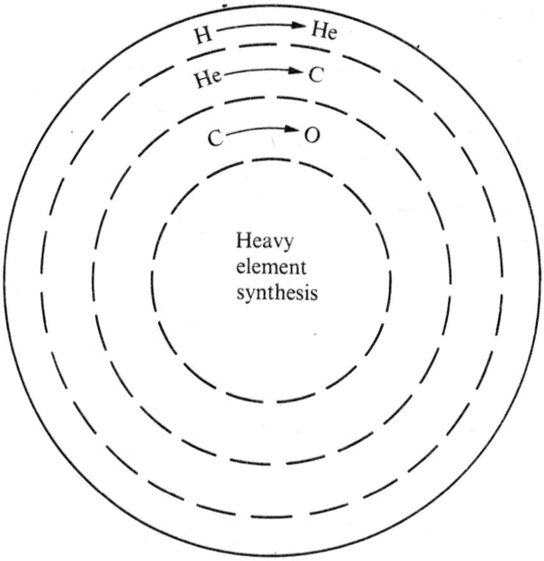

Fig. 1.2. Nucleosynthesis

different shells, see Fig. 1.2.

If the fusion takes place in a degenerate region of the star where the pressure is essentially decoupled from the temperature, then there can be no hydrodynamic adjustment to the rise in temperature resulting from the release of energy by the fusion reaction. In this event the temperature will rise locally extremely quickly, accelerating the rate of fusion reactions (fusion flash) until the material is no longer degenerate. Little of the energy from this localized explosion is radiated away from the star, however, most of it being absorbed in the outer regions of the stellar material.

As a source of energy, however, the fusion process ends with the formation of Fe^{56}. There are now no more exothermic fusion reactions to generate either radiation pressure or hydrodynamic pressure with which to withstand the gravitational forces, and the star will collapse until it is supported by the electron Fermi pressure. It is now a white-dwarf star, and as it radiates away its remaining free energy it cools down, and if equilibrium is achieved it finally becomes a black dwarf.

The Fermi pressure originates from the Pauli exclusion

principle which requires that no two identical fermions can be in the same state, and this means that even at the absolute zero of temperature there must be an occupation of non-zero momentum states up to some Fermi momentum. This momentum distribution gives rise to a pressure. For a non-interacting system at 0 K the Fermi momentum p_F is related to the number density n for spin-½ fermions by

$$p_F = h/2(3n/\pi)^{1/3}. \tag{1.17}$$

For a non-relativistic degenerate system the Fermi energy ε_F is

$$\varepsilon_F \equiv \frac{p_F^2}{2m} \gg kT \tag{1.18}$$

and the resulting Fermi pressure P_F is given by replacing the thermal energy kT in the expression for the gas pressure P_G by the Fermi energy

$$P_F = \frac{h^2}{8m\mu}\left(\frac{3n}{\pi}\right)^{2/3} \rho \tag{1.19}$$

Note that the Fermi pressure is inversely proportional to the mass m, and hence the lightest fermions (electrons) contribute most to the Fermi pressure, whereas most of the contribution to the mass density comes from the baryons

$$\rho \simeq n\mu \tag{1.20}$$

so that $P_F \simeq \rho^{5/3}$.

A scale of densities can be obtained by considering a system in which the mean spacing between the particles is of the order of the Compton wavelength λ for the particles

$$\lambda = \hbar/mc \tag{1.21}$$

which corresponds to a mass density ρ_o

$$\rho_o = \mu(\hbar/mc)^{-3} \simeq 3 \times 10^7 \text{ g cm}^{-3}. \tag{1.22}$$

At this density the electron Fermi momentum has become (1.17)

$$p_F \simeq mc \tag{1.23}$$

and hence the electrons are highly relativistic and the dispersion relation (1.18) should be replaced by

$$\varepsilon_F = \sqrt{(p_F^2c^2 + m^2c^4)} - mc^2. \tag{1.24}$$

At a density ρ_0, $\varepsilon_F \simeq mc^2 \simeq 0.5$ MeV and the system can be considered degenerate for temperatures $kT \ll \varepsilon_F$, i.e. $T \ll 10^9$ K. At very high densities we can approximate (1.24) by

$$\varepsilon_F \simeq p_F c \tag{1.25}$$

whence the Fermi pressure becomes

$$P_F = \frac{hc}{2\mu}\left(\frac{3n}{\pi}\right)^{1/3} \rho \sim \rho^{4/3}. \tag{1.26}$$

Inserting expressions (1.19) and (1.26) into our basic equilibrium equation (1.5), using $\rho = M/R^3$ and solving for M, we find that the mass of black dwarf stars is given by

$$M = \left(\frac{mc^2}{G\mu}\right)^{3/2} \sqrt{\rho/\rho_0}, \quad \rho \ll \rho_0 \tag{1.27}$$

and

$$M = M_c = \left(\frac{mc^2}{G\mu}\right)^{3/2} \frac{1}{\sqrt{\rho_0}}, \quad \rho \gg \rho_0 \tag{1.28}$$

i.e. for $\rho \gg \rho_0 M(\rho)$ becomes a constant and the Chandrasekhar mass is the upper mass limit for a stable, non-rotating black dwarf star. Solving instead for the radius we find

$$R = R_c(\rho_0/\rho)^{1/6} \tag{1.29}$$

where a suitable size scale is given by

$$R_c = \lambda\left(\frac{\hbar c}{G\mu^2}\right)^{1/2} \simeq 10^9 \text{ cm} \tag{1.30}$$

and we see that we have entered a region in which the radius decreases as the density (and hence the mass) increases.

The minimum mass of a black dwarf star or the maximum mass of a planet is given by (1.27)

$$M_P \simeq M_c(\rho_P/\rho_o)^{1/2} \tag{1.31}$$

where the planetary density is given by

$$\rho_P(P=0) \simeq \mu/a_o^3 \simeq 10 \text{ g cm}^{-3} \tag{1.32}$$

and a_o is the Bohr radius, thus

$$M_P \simeq 2 \times 10^{30} \text{ g}. \tag{1.33}$$

General relativistic effects for a black dwarf star are of order (eqn (1.10))

$$\delta \simeq \mathcal{R}/R \simeq \frac{m}{\mu}(\rho/\rho_o)^{2/3} \simeq 10^{-4} \tag{1.34}$$

In the white dwarf stage the star consists of positive nuclei (helium in the lightest stars, iron in the heaviest) permeated by a free relativistic gas of electrons. Since we are interested in neutron stars we shall follow the evolution of the more massive white dwarf stars. As a massive white dwarf star cools we surely have the ultimate in super metals. A predominantly ^{56}Fe lattice will evolve, with a few heavier elements like cobalt and nickel towards the centre, and lighter elements like chromium and manganese in the surface – in fact a chromium-plated stainless steel sphere! The whole star is permeated by a completely free degenerate electron gas. If the Fermi energy of the electron gas should exceed the neutron–proton mass difference, a new type of exothermic reaction becomes possible, viz. inverse beta-decay

$$e + p \rightarrow n + \nu_e. \tag{1.35}$$

The electrons are absorbed by the nuclei and convert protons into neutrons yielding progressively more and more neutron-rich nuclei. This process proceeds extremely rapidly because of the very high electron density and energy. Since the electrons are being absorbed their density falls and hence so does the Fermi pressure resulting in the renewal of the gravitational collapse. The energy released is carried away from the star by the neutrinos. The Fermi pressure of the electrons is now supplemented by the neutrino pressure in the same way that the hydrodynamic pressure in a main sequence star is supplemented by the radiation pressure. The dynamics of the collapse is obscured by the uncertainty in the opacity of dense matter to neutrinos. If the opacity is sufficiently great for the neutrinos from the interior to be unable to escape, their energy is redistributed in the stellar material with a resultant rise in temperature. In an extreme case this could lead to an explosion – a supernova? If the opacity is low and the neutrinos escape easily, then the star remains cold. Indeed, if the neutrino cooling effect is too great, the formation of elements heavier than helium may be inhibited in all but the most massive stars. We shall consider neutrino cooling in more detail in §3.1.

In an extremely over-simplified model the electron capture process would continue until all the electrons were absorbed and all the protons were converted into neutrons, and the nuclei would dissolve into a homogeneous neutron fluid. Such a star could be stabilized by the neutron Fermi pressure. In this case eqns (1.17) to (1.30) would be valid with the electron mass m replaced by the neutron mass $\sim \mu$. Note that the critical density analogous to ρ_0 would now be

$$\rho_N = \mu\left(\frac{\mu c}{\hbar}\right)^3 \simeq 2 \times 10^{17} \text{ g cm}^{-3} \tag{1.36}$$

while a star of the Chandrasekhar mass would be in equilibrium at a density (eqn (1.27))

$$\rho_C = \rho_N{}^2 M_C{}^2 \frac{G^3}{c^6} \simeq 10^{15} \text{g cm}^{-3} \tag{1.37}$$

and we note that $\rho_C \ll \rho_N$ so that the use of the nonrelativistic equation (1.27) is justified. The radius of a star of mass M_C and uniform density ρ_C would be

$$R_N \simeq 10 \text{ km} \tag{1.38}$$

and in this case general relativistic effects are of order

$$\delta = \mathcal{R}/R_N \simeq 0.3 \tag{1.39}$$

and should be taken into account.

A neutron gas at this density has a Fermi energy

$$\varepsilon_F \simeq 100 \text{ MeV} \tag{1.40}$$

and hence may be considered degenerate for temperatures less than $\sim 10^{12}$ K (eqn (1.18)). Within, at most, a few days of its formation neutrino cooling is expected to lower the temperature of a neutron star to less than 10^7 K.

The exact dynamical process by which the neutron star is formed is still obscure. One scenario envisages a giant star which has largely exhausted its nuclear fuel. The exhaustion of fuel will occur first in the core of the star, and over a period of thousands of years a massive core will evolve, supported principally by a sufficiently high temperature. As the core cools it becomes unstable against gravitational collapse. The inner regions of the core collapse fastest, and in a fraction of a second they reach nuclear densities where the collapse is halted by the nucleon degeneracy pressure. When the in-fall of this massive core is halted, the enormous kinetic energy is converted into heat. The temperature rises dramatically to more than 10^{12} K. As the temperature rises, so does the pressure. This explosion (supernova) drives off the slowly infalling envelope of the star producing a high flux of cosmic rays and an expanding ion cloud, e.g. the crab nebula. The explosion also compacts the core; and a neutron star is born. What fraction of the original stellar mass finishes up in the neutron star core is not clear and depends critically on the assumptions of the calculation.

We conclude this introductory section by summarizing the mass–radius relationships for various stellar objects in Fig. 1.3.

So far we have ignored the effects of rotations, which, as

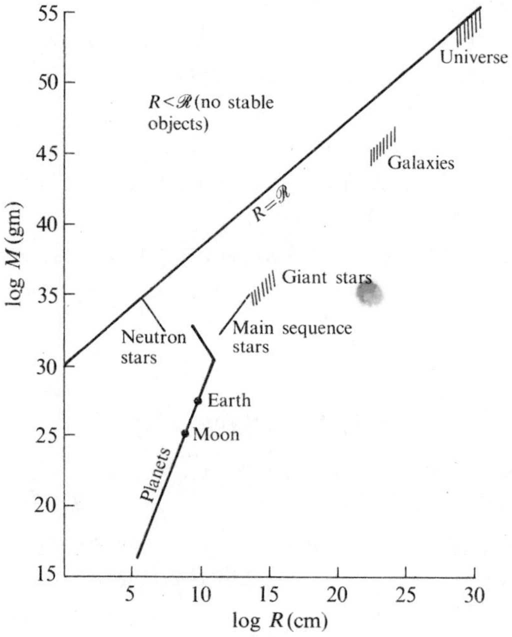

Fig. 1.3. Mass–radius relationships for various astrophysical objects.

we shall see, can be considerable. The period of the sun's rotation is ~25 days. If angular momentum were conserved, and we consider the sun to be rotating as a rigid body, then according to classical mechanics the period would be proportional to the square of the radius. Using eqns (1.16) and (1.38) as a guide we would be led to believe that if the sun were to contract to the size of a neutron star its period would fall to $\sim 2 \times 10^{-4}$ s. This is clearly an oversimplification, but it would imply that a point on the equator of the star would have a velocity which was a significant fraction of the velocity of light.

A full relativistic treatment of rotating stars is not yet

available. Kerr has given an exact solution to Einstein's equations which is appropriate to the region outside the event horizon for a spinning black hole. However, in general this solution will not fit smoothly to an exact internal solution for a rotating star. The internal solution for a star depends on the stress-energy tensor, i.e. the equation of state, for the stellar material. We shall briefly discuss how general relativistic effects change our view of the magnetosphere at the end of §4.1; in §6.2 we shall outline how, once we know the equation of state, we may take relativistic effects into account when calculating neutron star masses, and in Appendix C we have gathered together a few of the more important results of general relativity which are appropriate in the study of neutron stars.

1.2. Gross characteristics of neutron stars

We have seen that typically a star of the Chandrasekhar mass in the collapsed state will have a radius of ~ 10 km and a mean density ~ 10^{15} g cm^{-3}. This density is greater than the typical saturation density at the centre of a terrestrial atomic nucleus $\rho_o \simeq 2.8 \times 10^{14}$ g cm^{-3}. Smaller mass stars will have larger radii (eqn (1.29)) and hence be less dense.

Of course, the density will not be strictly uniform, but may be expected to increase towards the centre of the star. As the density of matter is varied, we are accustomed to its changing from one phase to another. In subsequent chapters we shall identify various density regions and discuss the nature of matter at these densities. At this stage we shall only comment on the oversimplified picture of the neutron star presented in the previous section.

The neutron is but one phase of the nucleon, the other phase being the proton, and the nucleons are but one phase of an infinite (?) number of baryon phases. The condition for thermal equilibrium of a phase mixture is equality of chemical potentials (Fermi energies) of the phases and a minimum of the Gibbs free energy with respect to the populations of the phases — the energies must, of course, include the rest masses of the phases.[1] Hence there will always be some protons present in the neutron gas (~ 1 per cent at ρ_0), and in order to maintain electrical neutrality there must be a like number of negatively-charged particles, the most likely of which is the electron. At low densities the neutrons and protons will tend to condense out and form nuclei, as they do in the white dwarf stage. At high densities, as the Fermi energy rises, it will be energetically favourable to produce new species of fermions. We expect muons to appear[2] when the neutron-proton chemical potential difference exceeds the muon rest mass energy ~ 105 MeV, which we expect to be the case at densities ~ 10^{15} g cm^{-3}. The production of hadrons is complicated by the role of the strong nuclear interactions, which we shall investigate in later chapters.

In the region where general relativistic effects are important the basic equation for stellar equilibrium (1.5) is replaced by the Tolman–Oppenheimer–Volkoff equation

$$\frac{dP(r)}{dr} = -G\frac{\{M(r) + 4\pi P(r)r^3/c^2\}(\rho(r) + P(r)/c^2)}{r(r - 2GM(r)/c^2)} \qquad (1.41)$$

for a non-rotating star (see Appendix C).

The gravitational potential energy of an electron at the surface of a neutron star of mass M_C (eqn (1.14))[3] and radius R_N (eqn (1.38)) is

$$(PE)_{grav} \simeq GM_C m/R_N \simeq 10 \text{ keV} \qquad (1.42)$$

i.e. the electron is bound gravitationally to the neutron star a thousand times more strongly than it is bound to the nucleus in a terrestrial hydrogen atom. Assuming an isothermal atmosphere we may estimate the scale height of the atmosphere for

---

[1]We shall use the convention throughout this book that if we use the term 'chemical potential' we shall imply inclusion of the rest mass, and the term 'Fermi energy' will not include the rest mass.

[2]Lepton number conservation is maintained by the simultaneous appearance of muon antineutrinos which are assumed to escape the star.

[3]It may turn out that the maximum mass of a neutron star is less than M_C (see §6.2).

particles of mass m.

$$h = kT/mg \tag{1.43}$$

where g is the surface gravitational acceleration

$$g \simeq GM_C/R_N^2 \simeq 10^{14}\ \text{cm s}^{-2}. \tag{1.44}$$

Thus, at a temperature of $\sim 10^7$ K, the scale height for electrons in the atmosphere is $\sim 10^4$ cm, and for protons and neutrons ~ 5 cm. Of course, we do not have a charge separation on this scale because, as we shall see shortly, the behaviour of charges at the neutron star surface is dominated by the electromagnetic field and not the gravitational field.

A particle orbiting a non-rotating star with angular momentum l sees an effective potential (including centrifugal effects, see Appendix C)

$$V(r) = -\frac{GM}{r} + \frac{l^2}{2r^2} - \frac{\mathcal{R}l^2}{r^3} \tag{1.45}$$

which differs from the Newtonian expression by the presence of the third term. In Newtonian gravitation stable circular orbits with arbitrarily small radii and arbitrarily large binding energies are possible as $l \to 0$. According to general relativity the smallest circular orbit has a radius

$$r_o = 3\mathcal{R} \tag{1.46}$$

(see Fig. 1.4), which is just about the radius of a neutron star and yields a maximum mass of a non-rotating neutron star of $\sim 10^{34}$ g.

Some stars are known to have magnetic fields. In the case of the sun this is ~ 10 G. If during gravitational collapse magnetic flux is conserved, in the neutron star phase the magnetic field would have grown inversely as the square of the radius, i.e. for the sun to $B \sim 10^{13}$ G. The energy density stored in the magnetic field is

$$\varepsilon = B^2/8\pi \tag{1.47}$$

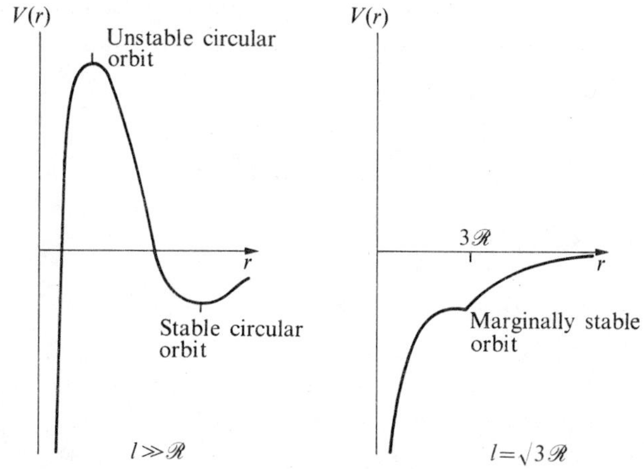

Fig. 1.4. The effective potential (eqn (1.45)) for various values of the angular momentum.

and for a field of 10^{12} G this is $\sim 10^{30}$ MeV cm^{-3} or equivalently $\sim 10^3$ g cm^{-3}! We shall study the effect of such large magnetic fields on matter in Chapter 4.

If the neutron star has a magnetic dipole moment $\mathscr{M}$ and this is at an angle α to the axis of the rotation, then according to classical electromagnetic theory the star will lose energy by electromagnetic radiation at a rate

$$d\varepsilon/dt = -\frac{32\pi^2}{3c^3}\frac{\mathscr{M}^2\sin^2\alpha}{\tau^4} \tag{1.48}$$

where τ is the period of the rotation. This loss of energy is supplied by a slowing down of the star's rotational frequency and a consequent loss of rotational energy. For rotations of extremely short period there is an even greater rate of energy loss due to gravitational radiation which varies as τ^{-6}.

Given that the magnetic axis is not parallel to the axis of rotation, there will be a time variation of the magnetic field, and hence there is necessarily a non-vanishing electric field. In the star the Fermi energy of the electrons is so high that they are essentially free, and so we may treat the star as a perfect conductor. Hence it is reasonable to assume that in equilibrium the Lorentz force should vanish, and thus the electric field $\mathbf{E}$ is

$$\mathbf{E} = \frac{1}{c}[\mathbf{B} \wedge (\boldsymbol{\omega} \wedge \mathbf{r})] \tag{1.49}$$

whence

$$\mathbf{E}\cdot\mathbf{B} = 0 \tag{1.50}$$

inside the star. Assuming that immediately outside the star there is a vacuum, we may solve the Laplace equation for the electrostatic potential ϕ with the boundary conditions (1.49) and the continuity of the electrostatic potential on the spherical surface of the star to give

$$\phi \simeq -\frac{\mathcal{M}\omega R}{3cr^2}P_2(\cos\theta) \tag{1.51}$$

i.e. an electrostatic quadrupole field with polar angle θ measured relative to the magnetic axis, which would result in a surface charge layer on the star. The ratio of the electrostatic force to the gravitational force for a charged particle at the surface of the star is

$$\Delta = \frac{e\phi}{GMm/R^2} \simeq \frac{eB\omega R^2}{3cGMm} \tag{1.52}$$

which for a proton at the surface of a Chandrasekhar mass star with a field of $\sim 10^{12}$ G yields $\Delta \simeq 10^9$. Thus, the system is dynamically unstable and the electrostatic forces will drive charges into the vacuum and hence build up the magnetosphere.

Suggested references

CLAYTON, D. *Principles of stellar evolution and nucleosynthesis*. McGraw Hill London (1968). An extremely stimulating account of stellar evolution.

HARRISON, B.K., THORNE, K.S., WAKANO, M., and WHEELER, J. (1965). *Gravitation theory and gravitational collapse*, University of Chicago Press. A very complete account of gravitational collapse.

BAYM, G. and PETHIC, C. (1975). *Rev. Nuc. Sci.* **25**. This review of neutron star physics contains an extremely extensive list of references to original research literature.

SEXL, R.U. (1973). *Black-holes physics*. CERN Lectures: TH.1759 CERN. A beautiful account of the evolution and properties of gravitationally collapsed objects.

HAWKING, S.W. (1974). *Nature, Lond.* **248**, 30. Discusses the possibility that quantum effects may lead to radiation being emitted by black holes.

2

PULSARS

2.1. Pulsar characteristics

Twinkle, twinkle little star,
How I wonder what you are.

MOST stars, of course, 'twinkle' because they are viewed through the Earth's atmosphere. However, extraterrestrial causes of stellar scintillation are also known. Thus, for example, the scintillation of radio signals from quasars is caused largely by irregular diffraction in the ionized solar atmosphere. There are also stars which are variable sources of energy, sending out pulses on regular periods measured in hours. We note that this is a period scale characteristic of planets or main sequence stars, see Table 1.1.

In 1967 a study of quasar scintillation was undertaken at Cambridge. In the case of regularly varying signals in a noisy background the sensitivity of measurements can be improved by using a long integration time, thus the regular signals can add coherently and will stand out more clearly against the noise of the background. Another critical feature in the Cambridge study was the use of a receiver with an extremely short response time. The maximum scintillation was expected when the radio source and the sun were close together, and no scintillation was expected when they were in opposite halves of the sky. In November 1967 the first pulsar was discovered as a fluctuating radio source observed crossing the meridian at night.

The characteristic signal from a pulsar has a pulse period of order a second; the shortest observed period is ~ 0.03 s and the longest ~ 4 s. The length of the pulse is of the order of a millisecond, indicating that the radio source is extremely small. the time of arrival of the pulse varies with the position of the Earth in its orbit around the sun, indicating that the source lies far outside the solar system, see Fig. 2.1. The orbital velocity of the Earth is $\sim 10^{-4}\,c$ and hence the

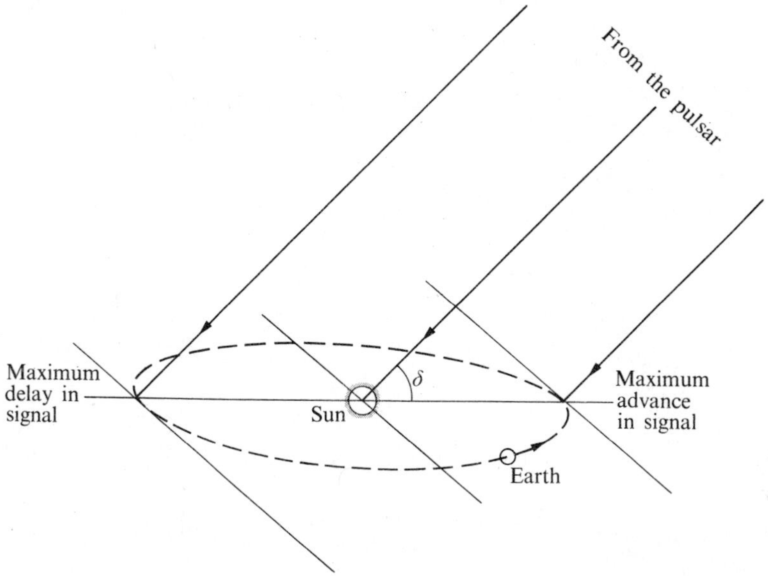

Fig. 2.1. Annual variation in pulse arrival time due to orbital motion of the earth. If δ is the elliptic latitude of the pulsar the amplitude of the variation is $\sim 500 \cos \delta$ s.

apparent periodicity of the pulsar can vary by one part in 10^4 due to the Earth's motion. When the Doppler variations due to Earth's motion around the sun are removed the pulse periods are generally stable to of order one part in 10^8, which is comparable with the stability of a quartz crystal oscillator. The general pattern is one of pulsar periods increasing by one part in 10^8 per year, so the period may be written

$$\tau = \tau_o \, e^{-t/t_o} \tag{2.1}$$

where the characteristic lifetime of the pulsar is

$$t_o \sim 10^6\text{–}10^8 \text{ years.} \tag{2.2}$$

Pulsars have been known to suddenly decrease their periods by factors of a few parts per million over a period of a few days and then to settle once more into a highly stable state. Such a sudden 'speed-up' is referred to as a 'glitch'. In

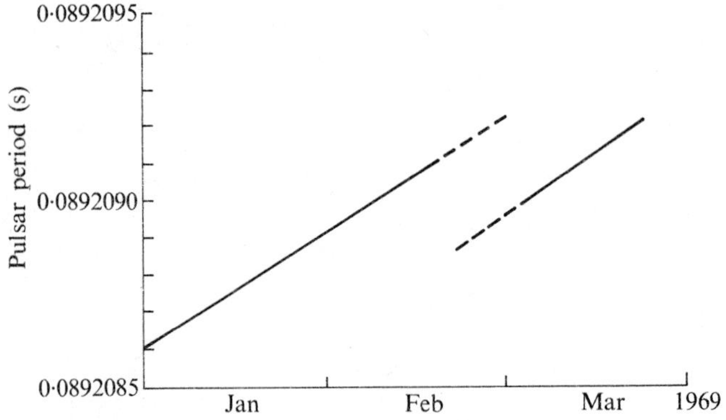

Fig. 2.2. The Vela 'glitch' of February 24 to March 3, 1969.

Fig. 2.2 we illustrate the famous 1969 'glitch' of the Vela pulsar PSR 0833-45.

All pulsars have been observed at radio frequencies. Many searches have been made in attempts to identify these radio sources by means of optically visible objects, but with very little success. Certainly the most successful and spectacular such identification is the case of PSR 0532, situated in the Crab nebula, which appears to radiate in the optical and x-ray regions of the spectrum as well as being a radio source. The emission from the Crab pulsar is pulsed at all frequencies with the same period to within a millisecond, see Fig. 2.3. The spectrum so far observed covers 39 octaves, ranging from radio frequencies $\sim$ 100 MHz to gamma quanta with energies in excess of 10 MeV (10^{16} MHz). The energy output of the Crab pulsar is greatest in the x-ray region, and the total measured electromagnetic energy output in pulses is $\sim 10^{30}$ W.

Pulsars are amongst the weakest observed radio sources, and hence measurement of individual pulses is difficult. However, when this can be achieved, the pulses show considerable structure, which can vary rapidly from pulse to pulse. Similarly, the polarization can vary not only from pulse to pulse but even in its distribution across the pulse itself. No identifiable correlations have been discovered for the variations in

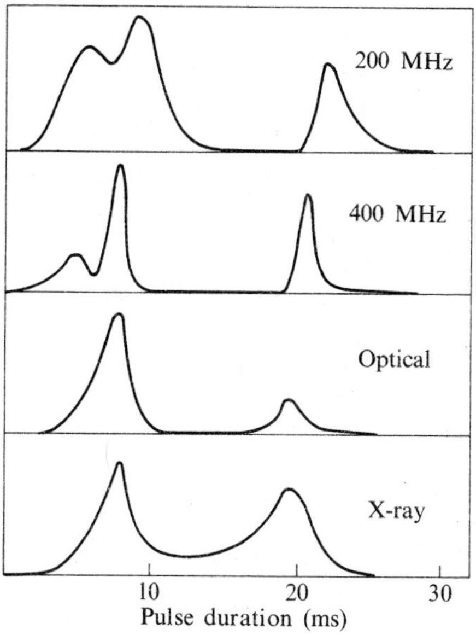

Fig. 2.3. Pulsed emission from the crab pulsar in the x-ray, optical, and radio spectrum.

the individual pulses, but the fact that there is such a variation should not be surprising. Apart from variations in the source, the signals reaching Earth will have been affected by dispersion and diffraction as they pass through the interstellar plasma, while polarizations will be affected by Faraday rotations produced by passage through magnetic fields, in particular the Galactic field.

If the interstellar medium is composed largely of ionized hydrogen, there will be an associated plasma frequency ν_p given by

$$\nu_p = \sqrt{(n_e e^2/m\pi)} \simeq \sqrt{n_e} \times 10^{-10} \text{ MHz} \tag{2.3}$$

where n_e is the electron number density. Clearly this plasma frequency is much smaller than the 100 MHz typical of radio signals. In the limit $\nu_p \ll \nu$ the group velocity of an electromagnetic wave through the plasma will be

$$v_g = c(1 - \nu_p{}^2/2\nu^2) \tag{2.4}$$

and it follows that the radio pulse is delayed compared with a free space signal by a time

$$t \simeq \int_{\substack{\text{signal}\\ \text{path}}} n_e(x)\,dx\ (e^2/2mc\pi)\nu^{-2}$$

$$\simeq 10^{-4} N_e \nu^{-2}\ \text{s} \tag{2.5}$$

where ν is in MHz and N_e is the total number of electrons per square centimeter along the line of sight. Thus, a receiver with a band width $\Delta\nu$ will have to accommodate a dispersion in arrival times Δt given by

$$\Delta t \simeq 2 \times 10^{-4} N_e \nu^{-3} \Delta\nu \tag{2.6}$$

which can be of the order of milliseconds and sufficient to wash out the fine structure of the pulse. As a result of passage of ionized dust clouds across the line of sight, N_e can vary significantly. While these might be expected to average out over large interstellar distances, this will not be the case for clouds associated with the source itself, e.g. in the Crab nebula.

Fluctuations in n_e will cause diffraction leading to the signal which is received on Earth being composed of rays with significantly different optical paths and interfering with one another, leading to scintillation, see Fig. 2.4. The band width $\Delta\lambda$ of spectral variations is given by

$$\Delta\lambda/\lambda \simeq \lambda/z\theta_s{}^2 \tag{2.7}$$

while the angle θ_s is given by

$$\theta_s \simeq (d/a)^{1/2} \Delta n_e r_o \lambda^2 \tag{2.8}$$

where d is the thickness of the diffracting screen, a is the scale of the irregularities in n_e, and r_o is the classical

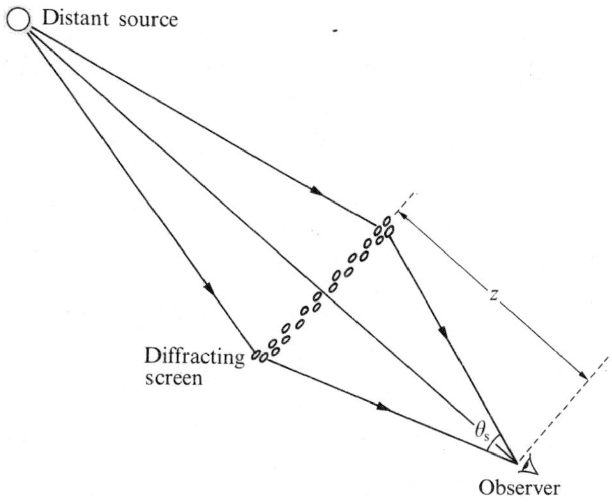

Fig. 2.4. A simple model for interstellar scintillation where the irregularities in the interstellar gas which lead to diffraction of the pulsar signal are assumed concentrated in a thin screen at a distance z from the observer.

electron radius. Typical results yield

$$a \simeq 10^4 \text{ cm} \qquad \Delta n_e \simeq 5 \times 10^{-5} \text{ cm}^{-3}. \tag{2.9}$$

Measurements of both scintillation and dispersion show that

$$\Delta n_e \propto n_e \tag{2.10}$$

making the interpretation of distance estimates from dispersion measurements uncertain.

The speed of fading depends on the velocities of the source and the observer relative to the interstellar diffracting medium. By measuring the delay as the interference pattern drifted between radio telescopes in England and Canada 6000 miles apart, estimates have been obtained which indicate that pulsars are moving with high velocities $\gtrsim 10^2$ km s^{-1}.

The refractive index of a plasma in a weak magnetic field H depends on the polarization. The plane of polarization of a radio wave of wavelength λ travelling through such a plasma rotates through an angle

$$\theta = R\lambda^2 \tag{2.11}$$

where the rotation measure is

$$R \simeq 2.5 \times 10^8 \int n_e(x)\mathbf{H}(x).\mathrm{d}x \tag{2.12}$$

with **H** in *G* and the wavelength λ and distance *x* in centimetres. For a receiver of bandwidth Δν in MHz the spread in polarization angle will be

$$\Delta\theta \simeq 18 \times 10^{20} \times R\ \Delta\nu/\nu^3. \tag{2.13}$$

Individual pulses are usually strongly ellipitically polarized and in a few cases even 100 per cent plane polarized. The degree of polarization is reduced somewhat in the integrated pulse. No clear picture of the fluctuations in polarizations has yet emerged.

The pulsars show a definite disc-like population with a concentration in the plane of the galaxy. The youngest pulsars, that is those of shortest period, for example the Crab pulsar, appear to be identified with supernova. There is no such obvious relationship for older pulsars. This might be expected since pulsars appear to move with velocities $\gtrsim 10^2$ km s^{-1}, and thus over a period $\sim 10^6$ years they will travel a distance $\gtrsim 10^{15}$ km, which would carry them clear away from the remnants of any supernova explosion. Further evidence for the high velocity of pulsars comes from their galactic distribution, which exhibits a scale height off the galactic disc $\sim 4.5 \times 10^{15}$ km, which is greater than that for any other class of objects in the galaxy except high velocity 'run away' stars.

2.2. Pulsars as neutron stars

The extreme precision of pulsar clocks and the extremely short period of the pulses excludes any known stellar object other than a neutron star from being a pulsar. All attempts to describe pulsars as vibrating white dwarfs, close binary white dwarfs, or planetary systems lead to periods at least an order of magnitude greater than those observed. Gold was the first to suggest that pulsars were single rotating neutron stars, and this suggestion was quickly followed by the prediction that the source of the pulsating signal could have its origins in the rapidly rotating magnetic dipole field associated with the star.

We have seen that a neutron star rotating with a period τ and having a magnetic dipole moment $\mathscr{M}$ at an angle α to the axis of rotation will radiate electromagnetic energy at a rate (eqn. (1.49))

$$\frac{d\varepsilon}{dt} = -\frac{32\pi^2}{3c^3}\frac{\mathscr{M}^2\sin^2\alpha}{\tau^4} \tag{2.14}$$

If this radiation is at the expense of rotational kinetic energy we can calculate the rate at which the period of the neutron star should increase. The rate of loss of rotational energy is

$$\frac{d\varepsilon}{dt} = -\frac{4\pi^2}{\tau^3} I \frac{d\tau}{dt}. \tag{2.15}$$

where I is the moment of inertia of the star. Using the charasteric mass $\sim 10^{33}$ g and radius $\sim 10^6$ cm we estimate that the moment of inertia $I \sim 10^{45}$ g cm^2. Equating the rate at which radiation is emitted to the rate of loss of rotational energy we find

$$\tau\frac{d\tau}{dt} = \frac{8\mathscr{M}^2\sin^2\alpha}{3Ic^3} \tag{2.16}$$

The smallest magnetic moment occurs for $\alpha = \pi/2$, and the corresponding magnetic field strength $B \simeq \mathscr{M}/R^3$. From the measured periods and rates of slowing down of pulsars we estimate that typically $B \simeq 10^{12}$ G. Due to the faintness of the pulsar signals it is unlikely that pulsars with $B \lesssim 10^{10}$ G could be observed.

If we calculate the second derivative of the period with respect to time, we find that

$$\frac{\tau\ddot{\tau}}{(\dot{\tau})^2} = -1 \tag{2.17}$$

independent of the particular mass or magnetic moment of the star. Hence the relationship (2.17) is a test of the basic hypothesis that the pulsar is a rotating neutron star with a large magnetic dipole moment. The only pulsar for which it is feasible to measure the second derivative of the period is that in the Crab nebula, in which case there is a discrepancy in the relationship (2.17) of the order of 10–20 per cent. That there is such a discrepancy is hardly surprising since the underlying dipole field may be expected to couple strongly to the plasma surrounding the star. This coupling is complicated by the possible charge separations which may be produced by the gravitational and induced electrical fields; to which must be added the complication of the general relativistic twisting of the magnetic field due to the Thirring–Lense effect (Appendix C). The resulting signal measured on Earth will be the results of the self-consistent interaction of the star with its magnetosphere and on the detailed model of the pulsing mechanism. No such fully self-consistent study has yet been completed.

Despite these uncertainties, there is no shortage of models to explain the pulsed nature of the signals. The most popular models involve a lighthouse effect, either with synchrotron radiation as a basis, or the relativistic beaming of radiation from an isotropic source, or some form of maser action in the magnetosphere. We shall examine some of these models in more detail in chapter 4.

While the regular pulsar signal may allow us to decide how the neutron star magnetic moment couples to the magnetosphere, it offers little information about the internal structure of the neutron star. To gain insight into this question we now consider the origin of glitches. First let us examine the nature of observed glitches (e.g. Fig. 2.2) in some more detail. There is, first, a sudden decrease in the period of the pulsar, such as we would associate with a sudden increase in the rotational velocity of the neutron star. This is followed by a relaxation time (a few days for the Crab pulsar, of the order of a year for the Vela pulsar) during which time the pulsar period stabilizes again. During this stabilization period the pulse rate can fluctuate corresponding either to 'speed ups' or 'slow downs' in the rotational velocity, but on a much smaller scale than that of the parent or 'macroglitch'. These smaller fluctuations in the pulsar period following a macroglitch we shall refer to as 'microglitches'.

A number of models have been proposed for glitch phenomena. These include:

(i) planetary perturbations;
(ii) magnetospheric instabilities;
(iii) accretion of matter;
(iv) hydrodynamic instabilities associated with the nucleon fluid;
(v) crustquakes; and
(vi) corequakes.

Initially the macroglitch phenomena could be explained in terms of the perturbative effects of a system of planets orbiting a neutron star. However, as sequences of macroglitches have been observed, and as the details of the structure of the microglitch relaxation process have been more closely studied, so more and more planets have had to be postulated in order to maintain the fits to the data. No planetary passage hypothesis is currentlv consistent with the detailed glitch phenomena observed.

Not all pulsars exhibit glitches; those in favour of attributing this phenomena to magnetosphere instabilities have to answer the question, why are some pulsar magnetospheres unstable while others are not? Observations of the Crab and Vela pulsars suggest that a major part of the magnetosphere must be blown away in each macroglitch. If this is the case, why is there no change in the pulse shape following a macroglitch? The characteristic time for the magnetosphere to fill is of the order of seconds, yet the period between macroglitches is of the order of years. Why does the instability take so long to develop?

Perhaps the most easily visualized model for macroglitch phenomena is that in which angular momentum is transferred to the star by infalling matter, Fig. 2.5. The first problem to discuss is the source of the accreting matter. To account for the observations one needs an accretion of $\sim 10^{23}$ g per macroglitch for the Crab pulsar, and $\sim 10^{26}$ g per macroglitch for the Vela pulsar. A neutron star which forms a compact x-ray source can only achieve accretion rates of $\sim 10^{24}$ g per year

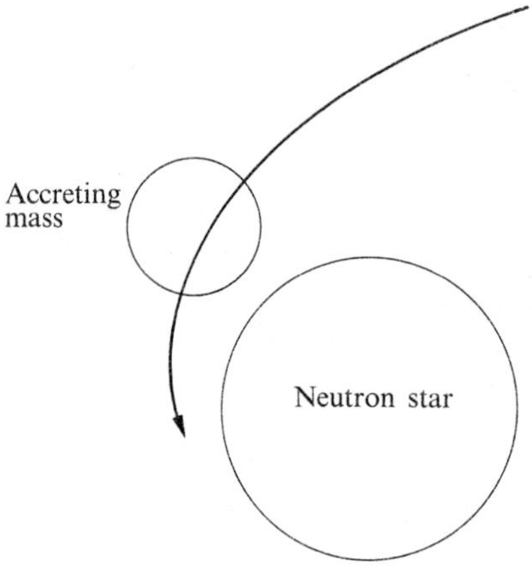

Fig. 2.5. Mass falling onto a neutron star can either lead to a 'speed up' or a 'slow down' in the rate of rotation of the star.

if it is postulated that there is a close companion star. The stability of the pulsar period argues against the existence of a close companion in the case of the Crab and Vela pulsars. Börner and Cohen have suggested that the matter accreted might be the 'fall back' of matter initially flung off at an early stage in the pulsar's life, e.g. during a supernova explosion which resulted in the creation of a neutron star. Colgate has suggested that all such fall back should occur within the first year of the supernova. These arguments cannot yet be considered conclusive and we shall not pursue them here; rather we shall limit our discussion to an analysis of how matter might be accreted and whether an accretion mechanism can indeed explain the macroglitch phenomena.

Assume that there is a planetary mass $\sim 10^{24}$ g in the form of a homogeneous sphere of radius R_s and density ρ (~ 5 g cm^{-3}) and with a Young's modulus μ ($\sim 10^{12}$ dyn cm^{-2}). It will be ripped apart by tidal forces before it reaches the star's surface. The shear strain is equal to the tangent of the shear angle, and at a distance R between the centre of mass of the star and the centre of mass of the infalling planet the shear angle ϕ is

$$\phi \simeq \frac{GM_c R_s^2 \rho}{10\mu R^3} \qquad (2.18)$$

where we have assumed the mass of the star $\sim M_c$. Thus pieces of accreting matter will only be stable against tidal breakup provided

$$\phi \simeq 10^{14} R_s^2/R^3 \lesssim \phi_c \qquad (2.19)$$

where the critical shear angle ϕ_c for planetary matter is $\lesssim 10^{-4}$. At the surface of the star $R \sim 10^6$ cm and the radius of stable accreting pieces of matter will have to be $\lesssim 1$ cm. Thus, as the planet falls towards the star it will begin to break up. The fragments of the infalling planet will begin to move independently when the gravitational forces acting on them due to the presence of the star exceed the gravitational forces holding them together, i.e.

$$\frac{GM_c}{R^2} \gg G\rho R_s \qquad (2.20)$$

or

$$R \ll 10^{16}/\sqrt{R_s} \text{ cm} \qquad (2.21)$$

Thus, at a distance $\sim 10^{10}$ cm from the centre of the star, the fragments will begin to move independently of each other. The fragments will form a Saturn-like ring, because of their large angular momentum, and will fall to the surface as a result of atmospheric friction and tidal forces. It is difficult to see how such a process could lead to the sudden speed-up observed in a pulsar macroglitch. Finally, any substantial accretion of matter leading to a macroglitch should mean that the speed-up is immediately preceded by an x-ray pulse, which is not observed.

If the neutron star consists substantially of a nucleon fluid, as the star slows down the fluid may become classically unstable. The slowing-down torques will set up a rotational flow in the fluid in which the angular momentum per unit mass decreases with the distance from the rotation axis. Such an instability is well known in laboratory studies of classical fluids. On such a model the interval between glitches should be approximately the relaxation time for the fluid flow. However, the observed relaxation time is two orders of magnitude less than the period between macroglitches for the Crab pulsar, while in the case of the Vela pulsar the relaxation time is actually greater than the period between macroglitches, and in the slower pulsars no macroglitches have been observed. As we shall discuss in the later chapters, it is likely that the nucleon fluid is a superfluid, in which case this instability may not arise, because the external torque will have to vanish. A well-known laboratory superfluid instability in which crust-pinned vortex lines tear themselves loose has also been proposed in order to explain glitches. Again the discrepancy between the relaxation time and the period between macroglitches is a problem, in addition to which pinning should be easier in the slower pulsars, where unfortunately no glitches are observed. Finally, for the estimated parameters of the superfluid and the structure of the crust (see later chapters) no pinning should occur, because the tension from a vortex bundle would be too great.

To conclude our discussion of possible sources of glitch phenomena we turn to starquakes. The existence of starquakes implies that some portion of the star is solid, and in Chapter 5 we shall investigate in some detail the possible nature of matter in various regions of the star. In order to make detailed comparisons between the elastic energy released in a starquake and the observed glitch phenomena, we need to calculate the amount of solid matter in the star, its distribution and density, and its elastic moduli. In other words, we require a detailed equation of state, and this problem will be examined in Chapters 5 and 6. For the present we shall summarize our results. The low density ($\lesssim 10^{14}$ g cm^{-3}) crust will be a nuclear lattice of which black dwarf matter is the prototype. The nuclear matter density region $10^{14} \lesssim \rho \lesssim 10^{15}$ g cm^{-3} is likely to be a nucleon superfluid, while the high density region $\rho \gtrsim 10^{15}$ g cm^{-3} may contain a solid core of hadronic matter. In extremely light neutron stars ($\sim 10^{32}$ g) most of the body of the star will be composed of the nuclear solid crust material, with only a small amount of nucleon superfluid in the centre. In medium mass neutron stars ($\sim 10^{33}$ g) the crust, nucleon superfluid, and core material will all be present; the proportion of each depending on the mass of the star. In these medium mass neutron stars most of the matter is in the nucleon superfluid phase with crust thicknesses $\sim 10^2$ metres. As the mass of the star grows, the core region rapidly grows and various new phases of superdense matter become possible.

Consider a rotating solid which would be spherical and have a moment of inertia I_o were it not rotating. Then, because it is rotating, it will be oblate and described by a deformation parameter ε such that the moment of inertia is

$$I = I_o(1 + \varepsilon). \tag{2.22}$$

The deformation parameter ε is a function of angular velocity, the gravitational acceleration due to the star and the elastic moduli of the solid. For the case of a neutron star which is slowing down, ε will be a function of time. The mechanical energy of the rotating solid is then

$$E = E_o - \frac{L^2}{2I_o}\varepsilon + A\varepsilon^2 + B(\varepsilon_o - \varepsilon)^2 \tag{2.23}$$

where L is the angular momentum, A measures the gravitational energy stored in the solid because of its rotation and has the value

$$A = 3/25\ GM^2/R \tag{2.24}$$

for an incompressible, homogeneous sphere of radius R, while B measures the elastic energy stored in the solid due to its distortion under rotation and is $\sim \mu V$ where V is the volume of the solid and μ is the shear modulus. ε_o is a reference distortion, which changes only because of plastic flow or a starquake, but is independent of the rotation. To find the equilibrium distortion we minimize the expression (2.23) for the mechanical energy for a fixed angular momentum $I_o\omega$, whence

$$\varepsilon = \frac{I_o\omega^2}{4(A+B)} + \frac{B}{A+B}\varepsilon_o. \tag{2.25}$$

In the crust material all neutron star models yield the result that the induced gravitational energy far exceeds the stored elastic energy, i.e. $A \gg B$ and hence

$$\varepsilon \simeq I_o\omega^2/4A. \tag{2.26}$$

There is much less agreement on the situation with regard to the solid hadronic core matter where predictions range all the way from a liquid spin crystal $A \gg B$ to an extremely rigid solid with $A \ll B$. Since in the lighter neutron stars the only solid is the crust, let us consider this situation first. The equilibrium shape is then dictated by eqn (2.26), and we see that as the angular velocity ω increases so also does ε, and thus the elastic energy E_{el}

$$E_{el} = B(\varepsilon_o - \varepsilon)^2 \tag{2.27}$$

also changes. When the mean stress σ exceeds a critical value σ_c, i.e.

$$\sigma = -\frac{1}{V}\frac{\partial E_{el}}{\partial \varepsilon} \simeq \mu(\varepsilon_o - \varepsilon) \gtrsim \sigma_c \tag{2.28}$$

a crustquake will take place, and elastic and gravitational energy will be released and converted into rotational energy of the crust. Accompanying such a glitch there will be a sharp change $\Delta\varepsilon$ in the shape of the star and a sharp change $\Delta\varepsilon_o$ in the reference oblateness

$$\Delta\varepsilon = (B/A+B)\Delta\varepsilon_o \simeq (B/A)\Delta\varepsilon_o. \tag{2.29}$$

This is directly observable since

$$\Delta\varepsilon = \Delta I/I = -\Delta\omega_\infty/\omega. \tag{2.30}$$

Thus, by measuring the change in frequency $\Delta\omega_\infty$ associated with the macroglitch, i.e. the difference in the stable angular frequency before the macroglitch and the stable angular frequency following complete relaxation after the macroglitch, and comparing this with the stable frequency prior to the macroglitch, we can measure the change in oblateness. From eqn (2.26) we estimate $\varepsilon = 10^{-4}$ while from macroglitch observations on the Crab pulsar eqn (1.20) would predict $\Delta\varepsilon \lesssim 10^{-8}$, i.e. the oblateness changed by 1 part in 10^4 during the Crab macroglitch.

After a starquake the star continues to slow down in a regular fashion (Fig. 2.2), and the stress will again build up until the next starquake occurs. The time t_{sq} between starquakes is given by eqn (2.26)

$$t_{sq} = \frac{2A}{I_o}\frac{\Delta\varepsilon}{\omega\dot{\omega}}$$

$$= \left(\frac{2A^2}{BI}\right)\frac{\Delta\omega_\infty}{\omega^2\dot{\omega}}. \qquad (2.31)$$

A numerical estimate of t_{sq} for comparison with experiment requires an ability to calculate the quantity $(2A^2/BI)$ which in turn requires a detailed equation of state for the neutron star matter. However, the coefficient A increases as M^2, while the moment of inertia is proportional to M. The crystal volume (and hence B) may also increase with mass, but is unlikely to increase faster than M^2. Thus, the time between starquakes is an increasing function of mass. Unfortunately most microscopic models yield an extreme sensitivity on the mass suggesting that for a Chandrasekhar mass star the time between macroglitches is $\sim 10^4$ years, while for a star of mass 1/10 M_c it is of the order of days. These have to be compared with time intervals of the order of years for observed pulsars. Such estimates suggest that the Crab pulsar has a mass $\sim 2 \times 10^{32}$ gm, while the Vela pulsar is two to three times more massive.

If the starquake occurs in the core, there is a greater uncertainty in the consequences, principally due to uncertainty in the elastic modulus of the core material. Some microscopic models suggest $\mu_{core} \gg \mu_{crust}$, in which case there is an enormous amount of elastic energy stored in the core and relatively small quakes with

$$(\Delta\varepsilon/\varepsilon) \sim 10^{-6} \tag{2.32}$$

can account for the observed macroglitches. There is no problem whatsoever in explaining the interval between macroglitches. Most models suggest the corequakes are more likely to occur in the massive and older neutron stars, while crustquakes are more probable in newer and less massive neutron stars.

All the macroglitch models involving accretion or starquakes attribute the relaxation and macroglitch phenomena to the transfer of angular momentum from the crust to the core (if it exists) and vice versa via the nucleon fluid. A study of the relaxation following a macroglitch can therefore lead to predictions concerning the amount and nature of the nucleon fluid within the neutron star.

We shall consider an extremely simple two-component model of a neutron star to illustrate the kind of predictions that can be made. One component is conducting and is assumed to rotate at the characteristic pulsar frequency ω, since all the charged particles are assumed to be strongly coupled to the magnetic field. This is a reasonble assumption since the conducting components' relaxation time, dictated by the Alfvén velocities, is very much shorter than the macroglitch relaxation time. Note that, although in different regions of the star some of the current carriers may be in a superconducting phase, there is no Meissner effect for neutron star matter. The moment of inertia of the conducting component of the star we shall denote by I_c. The other component is a neutral fluid (presumably mostly neutrons) having a moment of inertia I_n and characterized by a single rotational frequency ω_n, which couples weakly to the charged components. The weak coupling is described by

$$\begin{aligned} I_c\omega &= -\alpha - \frac{I_c}{\tau_c}(\omega-\omega_n) \\ I_n\omega_n &= \frac{I_c}{\tau_c}(\omega-\omega_n). \end{aligned} \tag{2.33}$$

The parameters α (the external rorque) and τ_c (the charged component relaxation time) will therefore depend on ω. Assume that there is a macroglitch at time $t = 0$ when there is a sudden jump $\Delta\omega$ and $\Delta\omega_n$ in the respective rotational frequencies; then, at a time t after the macroglitch, the angular frequency $\omega(t)$ will be given by

$$\omega(t) = \omega_o(t) + \Delta\omega\{Q\exp[-t/\tau] + (1-Q)\} \tag{2.34}$$

where $\omega_o(t)$ is the extrapolated angular frequency which would have pertained had there been no macroglitch and

$$Q = \frac{I_n}{I}(1 - \Delta\omega_n/\Delta\omega) \tag{2.35}$$

where

$$\tau = \tau_c \, I_n/I \tag{2.36}$$

with I the total stellar moment of inertia. The expression (2.34) (with different values of Q and τ) is in accord with the observations on the Crab and Vela pulsars.

Provided $I_c \ll I_n$ and/or $\Delta I_n \ll \Delta I_c$ we have

$$Q \simeq I_n/I \tag{2.37}$$

and Q can be obtained from (see eqn (2.31))

$$\Delta I/I = -\frac{\Delta\omega_\infty}{\omega} = -(1-Q)\Delta\omega/\omega \tag{2.38}$$

whence for the lighter neutron stars, e.g. the Crab pulsar, we find that 90–95 per cent of the star is in the form of a nucleon fluid, while for heavier neutron stars, e.g. the Vela pulsar, only 15 per cent of the stellar material is in the form of a nucleon fluid.

The coupling between the two components is characterized by the relaxation time τ. If the nucleon fluid were a normal Fermi fluid, the coupling would be so strong that the relaxation time would be $\sim 10^{-4}$ s while the observed time for the Crab pulsar is $\sim$ 8 days and for the Vela pulsar $\gtrsim$ 1 year. If the fluid is a superfluid, all the circulation is confined to vortex cores, and the spin up of the superfluid occurs when angular momentum is transferred to the vortex cores. The coupling is principally between the electrons, which form a normal Fermi fluid, and the nucleons in the vortex cores. The most obvious coupling is between the protons and the electrons through the Coulomb interaction. However, the proton abundance is so small that the direct magnetic interaction of the electrons with the neutrons is just as effective in communicating angular momentum to the superfluid.

The total coupling is thus proportional to the number of nucleons in the vortex cores, which depends on the size of the vortices and their density, times the probability of a nucleon quasiparticle excitation. The probability of a quasiparticle excitation is determined by the Boltzmann factor $\exp[-\Delta E/kT]$. The nucleon excitations are quantized in a plane perpendicular to the vortex core axis, hence

$$\Delta E \simeq h^2/2M\xi^2 \equiv \pi^2\Delta^2/4\varepsilon_F \qquad (2.39)$$

where M is the nucleon mass, ε_F is the nucleon Fermi energy, Δ is the superfluid energy gap, and ξ is the superfluid coherence length.

$$\xi = hv_F/\pi\Delta \qquad (2.40)$$

with v_F the Fermi velocity $[\varepsilon_F = \frac{1}{2}Mv_F^2]$. At densities $\sim 10^{14}$ g cm^{-3}, the superfluid gap is $\sim$ 1–2 MeV, and the Fermi energy is $\gtrsim$ 50 MeV. Hence, for neutron star matter at temperatures $\sim 10^8$ K ($kT \sim 0.01$ MeV), we have the probability of a quasiparticle excitation governed by a factor

$$\exp(-\pi^2\Delta^2/4\varepsilon_F kT) \simeq 10^{-14}. \qquad (2.41)$$

Thus we see that the two-step process, in which the electrons couple to the protons which in turn couple to the neutrons, is negligible, since this involves two such exponential factors one for the protons and another for the neutrons.

Because the gap parameter and the Fermi energy are functions of the nucleon density we find that the frictional force acting on the vortex depends on the position of the vortex in the superfluid. We shall content ourselves with demonstrating that the minimum relaxation time for electron—neutron coupling can be of the order of the observed relaxation times in the Crab and Vela pulsars. From the linearized Boltzmann transport equation, we find that the coupling is inversely proportional to the relaxation time, i.e.

$$\frac{1}{\tau} \simeq N_\nu \frac{\xi^2}{R^2} \frac{3}{4} \pi^2 \alpha^2 g_n^2 \frac{\varepsilon_F \Delta}{E_F^2}\left(\frac{2Mc^2}{\varepsilon_F}\right)^{\frac{1}{2}} \frac{kT}{\hbar} \exp(-\pi^2\Delta^2/4\varepsilon_F kT). \qquad (2.42)$$

N_ν is the number of vortex lines per unit area in a plane perpendicular to the axis of rotation, and R is the radius of the neutron star, so that $N_\nu \frac{\xi^2}{R^2}$ is the fraction of the superfluid in the vortex cores, since the radius of the vortices is $\sim \xi$, $\alpha = e^2/\hbar c$ is the fine structure constant, and g_n is the neutron g-factor equal to -1.91, while E_F is the extremely relativistic electron Fermi energy.

As the angular frequency increases, the density of vortices also increases and if ω_{crit} is the critical frequency at which the vortices become close packed, then

$$N_\nu \, \xi^2/R^2 \simeq \omega/\omega_{crit}. \qquad (2.43)$$

Parameters appropriate to the Crab pulsar suggest $\omega_{crit}/\omega \sim 10^{+18}$. For a superfluid gap $\sim$ 2 MeV and an electron Fermi energy $\sim$ 100 MeV, we find τ of the order of months. Clearly, shorter τ's can be obtained with smaller pairing gaps, or higher temperatures, or vice versa. Indeed, we might expect younger neutron stars to have higher temperatures and hence shorter relaxation times than the colder, older neutron stars, cf. the Crab and Vela pulsars.

We also conclude that there can be no great degree of turbulence in the nucleon superfluid since such turbulence would greatly increase the amount of fluid in the vortex cores and hence much reduce the relaxation times.

Finally, we comment on the microglitch phenomena observed in the Crab and Vela pulsars. As we have noted above, the frictional forces are not uniform throughout the superfluid, and this can lead to oscillations in the superfluid which can act back on the crust giving a noisy background to the general relaxation following a macroglitch. Alternatively, the conducting components in the crust may take time to move so as to maximally relax the stellar magnetic field stresses. These magnetic field stresses may lead to rumblings in the crust, again adding noise to the pulsar period. Another possibility we would mention is that the rotational axis may not be aligned with the elastic reference axis, and this would give rise to angular stresses which would result in localized microquakes much smaller than those arising from the radial stresses which we have suggested might act as a source for macroglitches. This model suggests that there should be more noise before a macroglitch than after it and could account for the report that the last macroglitch in the Crab was preceded by an increase in period noise. Lastly, models which predict both a solid crust and a solid core for neutron stars will have to produce detailed relaxation patterns for the neutron superfluid for comparison with observation.

It is clear that buried in the glitch phenomena there is a vast amount of information about the internal structure of neutron stars.

Suggested references

HEWISH, A., BELL, S.J., PILKINGTON, J.D.H., SCOTT, P.F., and COLLINS, R.A. (1968). *Nature, Lond.* **219**, 709. Announces the discovery of pulsars.

GOLD, T. (1969). *Nature, Lond.* **221**, 25. Suggests that pulsars are rapidly rotating magnetic neutron stars.

SMITH, F.G. (1977). *Pulsars*, Cambridge University Press. Reviews the observational data on pulsars and the interpretations which can be made directly from them. Contains comprehensive tables of positions, periods and other characteristics of all pulsars known to date, with references.

TER HAAR, D. (1972). *Phys. Rep.* **3**, 57. A theoretician's review with particular emphasis on the electrodynamics of the magnetosphere.

FEIBELMAN, P.J. (1971). *Phys. Rev.* D **4**, 1589. Estimates the effect of neutron superfluidity on postmacroglitch relaxation times.

3

NEUTRON STAR TEMPERATURES

3.1. Neutron star cooling

WE have until now virtually ignored any consideration of the temperature distribution within a neutron star, having simply asserted that we expect it to be 'cold' in the sense that the thermal energy kT per particle is much less than the Fermi energy. We shall now develop this argument in some more detail.

Any process which leads to a net reduction in the 'random' kinetic energy per particle in the star is a cooling process. The fact that the star slows its rate of collective rotation to compensate for the magnetic dipole radiation as discussed in §2.2 does not imply cooling because there is no loss of 'random' kinetic energy. Indeed, the slowing down may generate internal friction within the star leading to a heating of the stellar material.

The cooling process results in a loss of internal energy E for the star which is described by the luminosity $L(E)$. The dominant contribution to energy loss from the star will arise from photons, represented by L_γ and from neutrinos, represented by L_ν. The major source of photons will be electron bremsstrahlung, and, if we assign to the electrons a temperature T_e, we shall have a photon luminosity

$$L_\gamma(T_e) = 4\pi\sigma R^2 T_e^{\,4} \tag{3.1}$$

corresponding to a black body radiation pattern.

It was suggested long before the discovery of pulsars that hot neutron stars could represent compact x-ray sources. However, subsequent occultation experiments on the principal x-ray source within the Crab nebula indicate a source size of ~ light year compared with the radius of ~ 10 km for the pulsar, assumed to be a neutron star. Indeed, none of the observed galactic x-ray sources show observed spectra consistent with the black body radiation expected from the extremely compact surface of a hot neutron star. We shall show that this is not suprising, since, at high temperatures $T \gtrsim 10^6$ K, neutrino cooling processes are expected to dominate, while at the lower temperature the luminosity has fallen to such an extent that the star is likely to be unobservable. Indeed, prior to the discovery of pulsars it was argued that such considerations would render neutron stars unobservable.

An indirect statement about the internal temperature of a neutron star is provided by those models which require a substantial nuclear superfluid component to explain macroglitch relaxation observations. For superfluid neutrons we require that the neutron temperature T_n be such that the thermal energy kT_n is much less than the superfluid energy gap Δ_n which is itself a function of density. At densities $\sim 2 \times 10^{14}$ g cm^{-3} this implies that $T_n \ll 10^{11}$ K.

It is well-known that there is currently a serious problem in matching the observed solar neutrino luminosity to models of the sun, and, hence, discussions of neutrino luminosities for neutron stars must be highly speculative. At best we shall present for consideration some of the features which we feel to be of greatest importance for such a discussion.

There are of course very many nuclear reactions leading to neutrino production, e.g.,

(i) Neutrino pair bremsstrahlung by nucleons

$$\begin{aligned} n + n &\to n + n + \nu + \bar{\nu} \\ n + p &\to n + p + \nu + \bar{\nu} \end{aligned} \tag{3.2}$$

(ii) The modified Urca[1] process

$$\begin{aligned} n + n &\to n + p + e^- + \nu_e \\ n + p + e^- &\to n + n + \nu_e \end{aligned} \tag{3.3}$$

(iii) Pionic absorption

---

[1]The original Urca process is

$$e^- + \text{Nucleus }(A,Z) \to \text{Nucleus }(A,Z\text{-}1) + \nu_e$$
$$\text{Nucleus }(A,Z\text{-}1) \to \text{Nucleus }(A,Z) + e^- + \nu_e$$

and represents an essentially infinite sink for electron energies. It takes its name from the infamous Urca casino in Brazil which was an infinite sink for its clients' money.

$$\pi^- + n \rightarrow n + e^- + \bar{\nu}_e$$
$$\pi^- + p \rightarrow n + \nu + \bar{\nu} \qquad (3.4)$$

etc., together with a host of lesser reactions, e.g. neutrino pair production by collective electron plasma excitations, photoneutrino reactions, pair annihilation, and neutrino synchroton loss. The relative importance of the various reactions depends on the environment in which they take place. To first order this is simply a dependence on the equation of state, i.e. the density and the temperature, however, as the density and temperature vary, phase transitions can occur in the stellar material (see Chapters 5 and 6), and particular reactions can be enhanced or subdued by the nature of the phases present. Only recently have calculations taken into account the modifications due to nuclear superfluidity, pion condensation, etc. The occurrence of reactions producing neutrinos is not sufficient to guarantee a cooling of the system; this will also depend on the ability of the neutrinos to escape from the star. Prior to 1974 and the discovery of weak neutral currents, it was confidently asserted that the mean free path of neutrinos in neutron star matter was many orders of magnitude greater than the stellar radius. The argument went as follows: the typical neutrino energies will be of order kT and hence are expected to be less than ~ 1 MeV (corresponding to $T \lesssim 10^{11}$ K). Reactions of the type

$$\begin{aligned}&\nu_e + n \rightarrow p + e^-\\ &\bar{\nu}_e + p \rightarrow n + e^+\\ &\bar{\nu}_e + p + n \rightarrow n + n + e^+\\ &\nu_e + n + n \rightarrow p + e + n'\end{aligned} \qquad (3.5)$$

etc. are forbidden in this energy range by conservation of energy and momentum, and hence the dominant source of neutrino energy loss, it is argued, will be due to scattering from electrons. For neutrinos of energy ε_ν propagating in a degenerate electron gas of Fermi energy ε_F ($\varepsilon_\nu \ll \varepsilon_F$) the neutrino–electron scattering cross-section is

$$\sigma \simeq (2 \times 10^{-44})(\varepsilon_\nu/m_e c^2)^2(\varepsilon_\nu/\varepsilon_F)\,\mathrm{cm}^2 \qquad (3.6)$$

where the scale $\sim 10^{-44}$ cm^2 is given by the observed inverse β-decay cross-section. For antineutrinos the cross-section (eqn 3.6) should be multiplied by 1/3. The mean free path for electron neutrinos is

$$\lambda_\nu = [\sigma n_e]^{-1} \tag{3.7}$$

where the electron density n_e is given by the Fermi energy ε_F and vice versa. Thus, the electron–neutrino mean free path is approximately given by

$$\lambda_\nu \simeq (5 \times 10^3)(\rho_o/\rho)^{4/3} 1/\varepsilon_\nu{}^3 \text{ (km)} \tag{3.8}$$

where ρ_o is the nuclear saturation density (2.8×10^{14} g cm^{-3}), and the neutrino energy ε_ν is measured in MeV. The muon–neutrino mean free path is even longer because of the relatively much lower abundance of muons.

The forbidden reactions (3.5) all include a baryonic charge exchange, because the only known weak interaction at that time was conventional beta-decay, which, since it involves electrons which are charged, requires a change in baryon charge in order to conserve total charge. In the late 1960s Weinberg and Salam suggested the existence of a weak neutral current which would allow neutrinos to couple to baryons without any charge exchange. This would allow, for example

$$\nu + n \to \nu' + n' \tag{3.9}$$

i.e. inelastic neutrino–neutron scattering.

In the limit of nonrelativistic neutrons we would have for the neutron–neutrino Fermi part of the neutral current coupling

$$H_{\nu n} = \frac{g}{2V^2} \int [\bar{\nu}\gamma_o(1-\gamma_5)\nu] n_n \, d\mathbf{r} \tag{3.10}$$

where g is the weak coupling constant of ordinary beta-decay, and $n_n(\mathbf{r})$ is the neutron number density, i.e., we may expect a contribution to the neutrino opacity from the scattering of neutrinos from neutron density fluctuations in the stellar material. Thus the differential cross-section for scattering through an angle θ with momentum transfer $\mathbf{q}$ is

$$\frac{d\sigma}{d(\cos\theta)} = \frac{\varepsilon_\nu^2}{8\pi} g^2(1 + \cos\theta)|\delta n_n(\mathbf{q})|^2 \tag{3.11}$$

where $n_n(\mathbf{q})$ is the Fourier transform of the neutron density. Fluctuations in the neutron density can arise from thermal fluctuations in the neutron fluid or the local clustering of the neutrons into nuclei. In the former case we shall assume that $|\delta n_n(\mathbf{q})|^2$ is given by the mean square density fluctuations which can be calculated from the equation of state

$$|\delta n_n(\mathbf{q})|^2 = kT\, n_n^2\, V/K_T \tag{3.12}$$

where V is the volume considered and K_T is the isothermal bulk modulus

$$K_T = -V(\partial P/\partial V)_T \tag{3.13}$$

For a degenerate Fermi gas of neutrons we arrive at a neutrino mean free path

$$\lambda_\nu = \frac{K_T}{\varepsilon_\nu^2 (kT) n_n^2} (3 \times 10^3) \text{ km} \tag{3.14}$$

where the neutrino energy ε_ν and thermal energy kT are to be measured in MeV, n_n is in neutrons fm^{-3} while K_T is in MeV fm^{-3}. The neutrino mean free path depends on the equation of state for the stellar material through the modulus K_T. For interacting neutron matter at nuclear matter densities (see Chapter 6) we find

$$K_T/n_n^2 \simeq 250 \text{ MeV fm}^3 \tag{3.15}$$

and assuming $\varepsilon_\nu \simeq kT$ then the mean free path only falls below 1 km for temperatures in excess of 10^{12} K. However, the mean free path may be sharply reduced if the equation of state softens as it does whenever there is a phase transition; i.e. K_T is undefined at a phase transition. Phase transitions of course depend on the interactions in the system, and, hence, estimates of neutrino opacities which ignore the interactions, e.g. neutron Fermi gas models, may be wrong by many orders of magnitude. Nevertheless, we can place limits on the total opacity due to any softening of the equation of state. Suppose there is a layer of thickness δR over which the nucleon density changes by $\delta\rho$. Then the pressure change

$$\delta P = K_T \, \delta\rho/\rho \tag{3.16}$$

must be balanced by the gravitational forces in order to preserve hydrodynamic stability, thus

$$K_T = \frac{M(R) G \rho^2 m_N}{R^2} \frac{\delta R}{\delta \rho} \tag{3.17}$$

The number N_λ of mean free paths in such a layer is

$$N_\lambda = \delta R/\lambda_\nu \tag{3.18}$$

and inserting eqns (3.14) and (3.17) into eqn (3.18) we conclude that N_λ can be as large as ~ 20. It should be stressed that this is a theoretical upper limit on the opacity and to date most nuclear fluid models for neutron stars would suggest a substantially lower opacity.

Next we turn briefly to the suggestion that the nucleons may group together to form nuclei. We have already suggested that such a clustering takes place in the formation of the solid crust of the neutron star which we shall study in detail in §5.2. We now return to eqn (3.11), but now $n(r)$ represents the density distribution of nucleons in the nucleus, and there is a statistical factor for the local neutron-to-proton ratio. At low neutrino energies ($\lesssim$ 100 MeV) the neutrino wavelength is much larger than the size of the nucleus which we can thus treat as a point and we will then have coherent neutrino–nucleus scattering for which

$$\frac{d\sigma}{d(\cos\theta)} \simeq a_o^2 \frac{g^2}{2\pi} \varepsilon_\nu^2 A^2 (1 + \cos\theta) \tag{3.19}$$

where current experiments suggest $a_o^2 = 0.2 \pm 0.1$, whence at a density $\rho \simeq 10^{12}$ g cm^{-3} with nuclei of mass number $A \simeq 50$, the the neutrino mean free path is only $\simeq$ 100 m, which is of the order of the thickness of the crust region for a neutron star of mass $\simeq M_c$. Thus in some stars the solid crust may be opaque to neutrinos, especially since in some models nuclei with mass numbers much larger than 50 are predicted and the opacity rises as A^2. Such processes may be relevant for the blowing off of a supernova mantle during the collapse from a white dwarf to a neutron star configuration.

Suppose that the neutron star is in an initial state ψ_i with energy E_i, and that, as a result of a neutrino-producing process, it enters a final state ψ_f with energy E_f, and that the change in energy is radiated away from the star in the form of the neutrino, then first-order perturbation theory and simple thermodynamics gives for the rate of energy loss

$$L_\nu = \frac{2\pi}{\hbar} \sum_{\nu} \sum_{f<i} |\langle \psi_f \nu | H_\omega | \psi_i \rangle|^2 \, \varepsilon_\nu \times$$

$$\times\ \delta(E_i - E_f - \varepsilon_\nu) \exp(-E_i/kT). \qquad (3.20)$$

Bearing in mind the cautionary comments made above we shall now assume that the neutron star is a uniform quasifree hadronic fluid in which case the matrix elements of the weak hamiltonian H_ω are easy to evaluate. We then find that the dominant contributions are from the Urca process

$$L_\nu \sim (10^{39}\ \text{erg s}^{-1})\ (M/M_c)\ (\rho_0/\rho)^3\ T_9^8, \qquad (3.21)$$

bremsstrahlung

$$L_\nu \sim (10^{35}\ \text{erg s}^{-1})\ (M/M_c)\ (\rho_0/\rho)^3\ T_9^8 \qquad (3.22)$$

and pionic absorption

$$L_\nu \sim (10^{45}\ \text{erg s}^{-1})\ (n_\pi/n_n)(M/M_c)\ T_9^6 \qquad (3.23)$$

where the subscript 9 on the temperatures implies that the temperatures are measured in units of 10^9 K. We shall argue in Chapter 5 that in massive neutron stars ($M \sim M_C$) we expect a pion condensation to occur in the core in which case the dominant cooling mechanism will arise from neutrino production following pion absorption.

Given the luminosity L we can define a cooling time t

$$t = \int_{E_i}^{E_f} dE/\bar{L}(E) \tag{3.24}$$

where $\bar{L}(E)$ is the average luminosity over the internal energy interval dE. For neutrino–pion cooling for a star $M \sim M_C$ and an initial temperature $\sim 10^9$ K we find a cooling time ($E_f = kT_f$, $T_f \lesssim 10^4$ K) of the order of days. Assuming the Urca process dominates in lower mass stars, the most pessimistic neutron star models suggest cooling times $\lesssim 10^3$ years after which time $T \lesssim 10^6$ K.

Assuming that we only have radiative energy transfer, and that the conductivity of the stellar material is so high that the only significant temperature gradients are in the surface of the star, the thermal structure will be governed by the three equilibrium equations

$$dT/dr = -\frac{1}{4\pi r^2}\frac{3}{ac}\frac{K\rho}{T^3}L(r) \tag{3.25}$$

$$\frac{dP}{dr} = \frac{G(P/c^2 + \rho)(4\pi r^3 P/c^2 + m)}{r(r - 2mG/c^2)} \tag{3.26}$$

$$\frac{dm}{dr} = 4\pi\rho r^2 \tag{3.27}$$

together with the equation of state

$$P = P(\rho, T) \tag{3.28}$$

where the total opacity K is

$$\frac{1}{K} = \frac{1}{K_R} + \frac{1}{K_C} \tag{3.29}$$

where K_R and K_C are respectively the radiative and conductive opacities.

Thus reliable calculations of the equation of state for

neutron star matter are of high priority for a wide range of problems and will be tackled in Chapters 5 and 6.

3.2. Superfluidity in neutron stars

It is well-known that systems of fermions in which there is an interaction which favours the coupling of pairs of particles in a particular two-body state can exhibit a phase transition to a superfluid (or, if charged, superconducting) state. Details of pairing superfluidity are presented in Appendix B. The prototype of this kind of phase transition is provided by the Bardeen–Cooper–Schrieffer (BCS) model of superconductivity. In the case of electrons in a metal the pairing is indirectly facilitated by the lattice, and it is the electron–phonon interaction which leads in second order to an effective electron–electron pairing. The polarization of the lattice is maximized when an electron with momentum $\mathbf{k}$ and spin σ pairs with an electron in the time-reversed state of momentum $-\mathbf{k}$ and spin $-\sigma$. Similar states also exist in nuclei where again the pairing is between the time-reversed states of angular momentum j with z-component m and j with z-component $-m$. In this case this is a direct result of the short range dominantly attractive s-state nucleon–nucleon interaction. In almost degenerate Fermi systems the pairing occurs dominantly between states at the Fermi surface, and, since in heavy nuclei (or neutron star matter) there is a considerable neutron excess, we need not consider neutron–proton pairing (time-reversed pair states do not coincide at the Fermi surfaces), and we need only consider neutron–neutron and proton–proton systems separately. Such pairing is responsible for a wide range of phenomena in nuclear structure physics including various odd–even mass effects and the reductions in the moments of inertia of deformed nuclei. Since at a phase transition the correlation length grows to infinity, we have to be careful about talking in terms of phase transitions in finite systems. Physically, what we are implying is that the pairing correlations have to be treated explicitly and are not simply reproduced by low order perturbation calculations.

We now consider to what extent pairing-induced superfluidity and superconductivity might be of importance in neutron star physics. The condition for a pairing phase transition is considerably more complicated than that there should be an interaction which is attractive for a particular pair of states, although this is clearly a necessary condition. It is also necessary that the temperature be low enough that the coherently paired state be macroscopically populated. This requires that the latent heat Δ associated with the phase transition (i.e. pairing gap) be much larger than the thermal energy kT. In turn the gap parameter Δ depends critically not only on the strength of the pairing interaction but also on the density of states at the Fermi surface. This introduces a complicated density dependence. First, the density of states at the Fermi surface depends on the density. For most quasiinfinite systems, and certainly for the Fermi gas, the density of states at the Fermi surface monotonically increases as the particle density increases, and hence we might expect the latent heat to increase. However, as the density increases so do the momenta of particles at the Fermi surface so one also has to consider the momentum dependence of the pairing interaction. Finally, it is not the bare fermion–fermion interaction which is significant but the effective interaction between quasiparticles at the Fermi surface which arises when all the renormalization effects of the many-body environment have been taken into account. These effects are particularly important in a strongly interacting system of nucleons and increase rapidly in complexity as the density increases.

Considering first the electrons, throughout the body of the neutron star the electrons are highly relativistic, and in the core and the crust† where there is the possibility of electron–phonon coupling this is so weak that it cannot give rise to a BCS type of pairing. Hence we do not expect electron superconductivity. In the neutron star 'atmosphere' we have to consider the properties of the polymeric solid discussed in §4.2. This is an insulator with regard to charge propagation perpendicular to the magnetic field lines, implying an extremely low density of states for electron modes propagating in this direction and hence no superconductivity. In the direction of the magnetic field lines, i.e. along the axis of the polymers, the magnetic field is not time reversal invariant and destroys any pairing that might otherwise occur and hence no BCS effect is to be expected.

We are thus left to consider the baryonic components of the star. At nuclear densities, i.e. $\rho \lesssim 2.8 \times 10^{14}$ g cm^{-3}, we know that both neutrons and protons have undergone a pairing transition in cold nuclear matter and that the pairing gap is 1–2 MeV. This suggests that for neutron star matter at temperatures $\lesssim 10^8$ K we expect to have superfluid neutrons and superconducting protons in regions of density 10^{14}–10^{15} g cm^{-3}.

In the crust of the star for densities 10^7 to $\sim 2 \times 10^{11}$ g cm^{-3}, the neutrons and protons condense out into nuclei which form a solid lattice. The nuclei in this lattice crust have densities similar to those found in terrestrial nuclei and hence are expected to be composed of superfluid neutrons and superconducting protons. As mentioned above we must be careful in discussing phase transitions in finite nuclei. What we are implying is that physically the pairing correlations have to be treated explicitly when describing the properties of nuclei forming the lattice crust. We have to be particularly careful when describing the excitation function for the nuclei in the crust. The pairing correlations have no major effect on the gross structure of matter in this region; the slight softening of the equation of state resulting from the lowering of the nuclear binding energy is on the level of 0.1 per cent and hence is negligible for most purposes.

At densities greater than $\sim 2 \times 10^{11}$ g cm^{-3} neutrons start to drip out of the nuclei. The nuclei themselves remain in a paired state since their density does not alter significantly. In the degenerate neutron gas the $^1S_0$ phase shift is negative, a necessary condition for *S*-state pairing; however, initially the density of states is so low that a phase transition is unlikely to occur. As the density rises so does the neutron-to-proton ratio in the nuclei and the ratio of neutrons in the gas to neutrons in the nuclei. At some point the neutron gas will become superfluid. The exact density at which this occurs is complicated by the fact that in addition to the direct $^1S_0$ interaction between the nucleons there is a weak coupling between the neutrons and the lattice phonons yielding a genuine

BCS-type tendency to pairing. Preliminary calculations suggest that an S-type pairing transition occurs at $\sim 10^{13}$ g cm^{-3}.

At densities around 10^{14} g cm^{-3} the nuclei start to dissolve and we have a degenerate fluid of neutrons and protons. The neutron fluid is superfluid, and represents ~ 90 per cent of the mass density in this region, and hence clearly dominates the structural properties of the star in this region. The proton gas is probably not superconducting, and there is, of course, a normal, degenerate, relativistic electron gas of number density equal to that of the proton gas preserving charge neutrality. The penetration of the normal fluid into the crust region suggests that there will be a strong coupling between the fluid and the crust.

At fermi momenta corresponding to densities $\sim 2 \times 10^{14}$ g cm^{-3} the $^1S_0$ phase shift becomes positive for the bare neutron—neutron interaction. Brueckner-type calculations suggest that in the many-body medium the disappearance of S-state pairing occurs at a slightly higher density $\sim 2.5 \times 10^{14}$ g cm^{-3}. The exact dependence of the predicted pairing gap on the density depends critically on the method of calculation and must be considered uncertain to ~ 15 per cent. At Fermi momenta corresponding to still higher densities the bare neutron—neutron $^3P_2$ phase shift which is negative becomes dominant. Pairing in this state would lead to anisotropic superfluidity analogous to that observed in the millidegree region for liquid helium-3. Such anisotropic pairing could have significant consequences for the rotational properties of the star, but this has not yet been investigated, largely because a number of other effects like neutron solidification and pion condensation are also likely to appear at such densities, and also because realistic calculations of the effective interactions at these densities are extremely uncertain at the present time.

Given the form of interaction between the particles we can calculate the pairing matrix elements (see Appendix B), and hence the pairing gap, and thus the superfluid critical temperature T_c. We have calculated the neutron superfluid critical temperature from the onset of the neutron drip region at $\rho \simeq 2 \times 10^{11}$ gm cm^{-3} up to the nuclear saturation density $\rho_o = 2.8 \times 10^{14}$ g cm^{-3} assuming that the neutrons interact with each other through the Skyrme interaction (see Appendix A). In Fig. 3.1 we indicate the density region for which the calculated critical temperature is greater than 10^9 K. We see that at the expected temperatures $T < 10^6$ K we are well below the

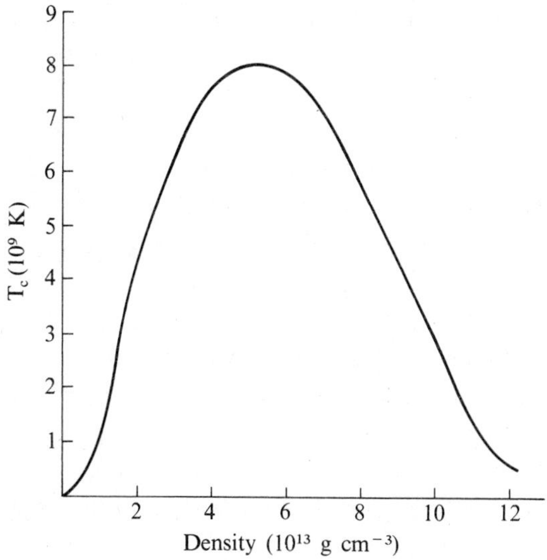

Fig. 3.1. Neutron superfluid critical temperature as a function of neutron star matter density.

critical temperature for most of this region and we would indeed expect the bulk of the neutrons to be superfluid. However, at any finite temperature even in a non-rotating superfluid, i.e. containing no vortices, some of the fluid will be normal (see Appendix B). The interaction of such a normal neutron fluid with the neutron vortex cores could seriously alter our estimates of the post-glitch relaxation times suggested in §3.2. If we repeat the above calculations for the protons, we find that throughout this region the proton density remains so low that superconductivity is never established. These normal protons can also interact with the neutron vortex cores and if we are to hold to our explanation of the long post-glitch relaxation times then this places severe constraints on the maximum permissible temperature.

Because of the presence of the electrons we do not expect to see a Meissner effect in the neutron star interior, i.e. the magnetic field will pervade the whole star, and, as a consequence, the charged components of the star will all be assumed to corotate. In addition it is assumed that the rotation of the charged particle system corresponds to the observed pulsar period. Information about the charged particle motion can only be communicated to the neutrons via interactions taking place between charged particles and neutrons in the vortex cores and we have discussed this in connection with the relaxation following macroglitches.

If we assume that the superfluid neutrons all rotate with a uniform angular velocity while the charges all corotate with the magnetic field, then because of the magnetic dipole radiation the pulsar (i.e. the charged particle component) slows down, and, hence, on average the superfluid is always rotating faster than the pulsar. Thus there will be frictional forces acting between the charged particle system and the neutron superfluid which can convert some of the rotational kinetic energy of the star into frictional heating. Hence, even though the star is near the endpoint of its evolution, there remains a source of internal energy. For our simplified model the rate of heat generation is given simply in terms of the pulsar period, the pulsar slowing down rate, the density of neutrons, the density of charges, and the macroglitch relaxation time which gives us information about the coupling between the charged components and the neutron superfluid. Such calculations establish that frictional heating cannot maintain surface temperatures in excess of $\sim 10^7$ K even in the fastest pulsars, and we remain firmly in the degenerate region and for the most part far below the critical temperature for pairing transitions.

We cannot really believe that the neutron superfluid will all corotate so conveniently. Certainly, because we expect a strong coupling between the neutron fluid and the nuclear lattice crust we expect the neutron fluid which permeates the crust to corotate with it. A rotating superfluid contains vortex lines. These are in a state of tension and will align themselves with the rotation axis. The vortex lines move essentially with the velocity of the neighbouring superfluid and hence if there is a velocity gradient in the superfluid the vortex lines will be distorted. Consider an isolated vortex line, and for the moment ignore the tension. The line will steadily stretch, slowly winding itself about the rotation axis until it is very tightly wrapped (see Fig. 3.2). Along the

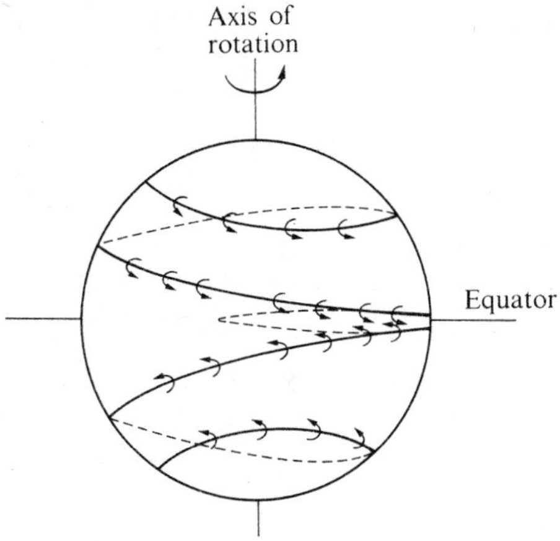

Fig. 3.2. A much elongated vortex line becomes wrapped around a rapidly rotating neutron star. Near the equator an unstable situation arises when sections of the vortex line of opposite sense are brought close together.

equator, lines of opposite sense are brought close together. This is unstable, and eventually the line will snap and reconnect to form a vortex ring. The process will repeat itself until the superfluid contains many vortex rings which interact with each other and the remaining distorted vortex lines; this is superfluid turbulence. Now, of course, there is tension in the vortex line, and hence there will be an angular velocity below which turbulence will not set in. So far we have discussed a single vortex line in isolation. The situation is complicated by the fact that we have an array of vortex lines. For an array of N closely-spaced lines the total differential force is N times greater than on a single line but this does not increase the probability of turbulence because the tension increases as N^2. It has been argued that, because of this, turbulence will not be established in neutron star fluid even for the fastest pulsars. A definitive answer to this question awaits developments in superfluid hydrodynamics.

Finally we consider the core of the neutron star at densities $\gtrsim 10^{15}$ g cm^{-3}. If the difference in the neutron and proton chemical potentials $\mu_n - \mu_p$ (including masses) exceeds the pion effective mass m_π^{eff} in dense matter then we expect the system to enter thermal equilibrium with respect to the process

$$n \to p + \pi^- \tag{3.30}$$

The pions being bosons can form a bose condensate provided the temperature is low enough. Also, since the dominant nucleon–pion interaction is in *p*-states it has the nonrelativistic form **σ**.**k** where **σ** is the nucleon spin and **k** is the pion momentum, and hence we expect the bose condensate to occur for a non-zero momentum state. A charged bose condensate of non-zero momentum is a supercurrent. We shall return to discuss this topic in §5.5.

Ultimately we should like to derive a self-consistent picture of neutron stars structure in which the various supercurrent components would act as the generators of the huge magnetic fields thought to be associated with pulsars. We are still some way from such a view at the present time.

Suggested references

BAHCALL, J.N. and WOLF, R.A. (1965). *Phys. Rev.* **B 140**, 1445 and 1452. The first paper deals with the properties of neutron stars at absolute zero temperature. The second paper deals with neutrino cooling prior to the discovery of neutral currents and the question of observability prior to the discovery of pulsars.

CHIU, H.Y. (1964). *Ann. Phys., N.Y.* **26**, 364. CHIU, H.Y. and SALPETER, E.E. (1964). *Phys. Rev. Lett.* **12**, 413. These papers deal with the photon emission from hot neutron stars and the question of observability.

FREEDMAN, D.Z. (1974). *Phys. Rev.* **D 9**, 1389. The first analysis of the inhibition of neutrino cooling as the result of the discovery of weak neutral currents.

KOGUT, J. and MANASSAH, J.T. (1972). *Phys. Lett.* **A 41**, 129. The first study of the enhancement of neutrino cooling rates due to the possible existence of a pion condensate.

YANG, C.H. and CLARK, J.W. (1971). *Nuc. Phys.* **A 174**, 49. CHAO, N.C., CLARK, J.W. and YANG, C.H. (1972). *Nuc. Phys.* **A 179**, 320. Calculate proton and neutron superfluidity critical points assuming $^1S_0$ state pairing.

RUDERMAN, M. (1967). In *Proceedings of the Fifth Annual Eastern Theoretical Physics Conference 1966* (ed. D. Feldman), Benjamin, New York. Raises the question of possible $^3P_2$ state pairing amongst neutrons at high density.

GREENSTEIN, G. (1974). In *Physics of dense matter, IAU Symposium No. 53, 1972* (ed. C. Hansen), D. Reidel, Boston. Reviews the effects of rotation

coupled with superfluid properties of neutron star interiors.
FEIBELMAN, P.J. (1971). *Phys. Rev.* **D 4**, 1589. Estimates the effect of neutron superfluidity on post-macroglitch relaxation times.

4

THE EXTERIOR OF NEUTRON STARS

4.1. The magnetosphere

IN our discussion of neutron star structure we have decided to order our material in accordance with distance from the centre of the star. If pulsars are indeed neutron stars then the pulsed signals recorded on earth may be considered as the asymptotic long range manifestation of neutron stars. We now move closer to the star, but not until the following section shall we begin to deal with the actual matter associated with the star. In this section we shall offer some thoughts on the nature of the fields in the vacuum immediately surrounding the star. For several reasons such a region really does not exist. Space is not a true vacuum and especially not in the vicinity of a star where matter may be being accreted onto, or ejected from the star, and such matter will interact with the fields surrounding the star. There is an additional complication in the case of young pulsars which may be formed as a result of supernova explosions, e.g. the Crab pulsar, in that the expanding cloud of matter which forms the nebula is linked with the source star. Of course, since apparently most pulsars are high speed objects, then pulsars older than $\sim 10^6$ years will have left the supernova clouds in which they were created, and the matter in the clouds will move independently of the fields generated within the neutron star.

With these considerations in mind we shall now proceed, and we shall begin by briefly commenting on the interaction between the electromagnetic field of the Crab pulsar and the expanding cloud of matter which is the Crab nebula. The first remarkable observation is that the expansion of the nebular material is still accelerating. Had the nebula been formed in a supernova explosion without any subsequent source of energy, we would have expected the nebular material to have received an initial outward velocity and, since the explosion, to have been decelerating under its own gravitational field. The fact that the nebula's expansion is still accelerating argues for a source of energy still driving the expansion. The mass of the nebula is uncertain but would appear to be in the range 10^{33}–10^{35} g, while the mass of the Crab pulsar is believed to be in the range 10^{32}–10^{33} g. We have suggested that pulsars may have velocities ~ 100 km s^{-1} and hence the Crab pulsar may have moved ~ 10^{12} km from the point in space where it was created. The present expansion velocity of the nebula is considerably greater than 100 km s^{-1} being of the order of 1000 km s^{-1} in the outer regions, and observations are not inconsistent with the Crab pulsar having been created at the centre of expansion of the nebula. Another argument in favour of a continuing source of energy within the nebula stems from the observed nebular x-ray spectrum. This, it is suggested, is synchrotron radiation arising from electrons with energies in excess of 10^{10} eV moving in magnetic fields ~ 10^{-4} G. Such electrons are extremely efficient radiators, and it is expected that they would lose most of their energy within ~ 100 years. The continuing existence of x-rays suggests that energy is being continually pumped into the electrons long after the supernova explosion.

Observations of the expansion of the nebula and estimates of the mass of nebular material suggest that ~ 10^{42} J of kinetic energy were released in the initial supernova explosion. This may represent only a fraction of the total energy released. The continued acceleration of the expansion and the x-ray observations suggest a continuing source of energy radiating 10^{31}–10^{32} W within the nebula. We note that the total energy content of the pulses observed from the Crab is ~ 1 per cent of this amount.

We now consider whether the magnetic dipole radiation from the Crab pulsar viewed as a rapidly rotating magnetic neutron star could be a candidate for the continuing source of energy within the Crab nebula. The rate of energy radiation is [eqn (2.14)].

$$\frac{d\varepsilon}{dt} = -\frac{32\pi^2}{3c^3}\frac{\mathcal{M}^2\sin^2\alpha}{\tau^4} \tag{4.1}$$

Assuming $\alpha = \frac{\pi}{2}$ and a surface magnetic field of 10^{12}–10^{13} G with $B \sim \mathcal{M}/R^3$ and a radius ~ 10^6 cm, we find, using the observed Crab pulsar period that

$$\left.\frac{dE}{dt}\right|_{\text{crab}} \sim 10^{31}\text{–}10^{32} \text{ W} \tag{4.2}$$

as required. Our final comment on this subject is that the x-ray observations suggest that the conversion of radiation energy into electron energy is remarkably efficient, the efficiency being $\gtrsim$ 10 per cent.

We now consider an 'isolated' magnetic neutron star and briefly comment on the region around the star in which the dominant phenomenon is the star's electromagnetic field. In this region there will be a low density plasma consisting of interstellar ions and matter ejected from the star itself – a subject to which we will return below. In this section we shall restrict our comments to the region where the behaviour of the plasma is dictated by the electromagnetic field but where the plasma has a negligible effect on the structure of that field. We would remind the reader that even the sun's magnetosphere is not well understood, and hence it should not come as a surprise that our comments on the neutron star magnetosphere are so severely limited.

We shall assume that the intrinsic field of the star is that of a magnetic dipole. If the suggestion that the slowing down of the pulsar is due to electromagnetic radiation from the rotating dipole is accepted, then the magnetic axis and the axis of rotation cannot be parallel. Thus there will be an induced electric field due to the motion of the neutron star

$$\mathbf{E} = \frac{1}{c}[\mathbf{V}+\boldsymbol{\omega}\wedge\mathbf{r}]\wedge\mathbf{B} \tag{4.3}$$

where $\mathbf{V}$, the translational velocity through space is $\sim$ 100 km s^{-1} and ω is the angular velocity of the star. The sun has a radius of $\sim 10^{11}$ cm and a rotational period of $\sim 10^{6}$ s. If we assume the sun rotates as a rigid body (it does not) and that angular momentum is conserved in the collapse to a neutron star which also rotates as a rigid body (it does not), then we would deduce that $\boldsymbol{\omega}\wedge\mathbf{r}$ for a point on the stellar equator would be $\sim 10^{10}$ cm s^{-1}, i.e. of the order of the velocity of light. Even if stars did rotate as rigid bodies this might be a poor estimate of $\boldsymbol{\omega}\wedge\mathbf{r}$ since if the neutron star were created in a supernova, as assumed in the case of the Crab, then the original angular momentum of the star will be shared by the neutron star, the nebular material, and, since the explosion appears to be asymmetric, (i.e. the neutron star is moving relative to the centre of expansion of the nebula) in the relative motion of the neutron star and the mass centre of the nebula. If we assume that ω is of the order of $10^3\ s^{-1}$ (see Table 1.1) then we conclude that $\boldsymbol{\omega}\wedge\mathbf{r}$ for a point on the equator is $\sim 10^9$ cm s^{-1} or $\sim$ 10 per cent of the velocity of light. Thus, while there is clearly some uncertainty, we would expect that $\boldsymbol{\omega}\wedge\mathbf{r} \gg \mathbf{V}$, and that $\boldsymbol{\omega}\wedge\mathbf{r}$ would be a considerable fraction of the velocity of light. If the star is a perfect conductor and is spherical, and if there is a vacuum outside the star, then the induced field will be an electric quadrupole field. However, as we shall see in the next section, the stellar 'atmosphere' immediately below the magnetosphere is likely to be a highly anisotropic conductor shielding the crust of the star (see §5.2) which is predicted to be a perfect isotropic conductor.

Theories of the structure of the electromagnetic field surrounding a neutron star are not well developed but most approaches begin by seeking the vacuum electromagnetic field and matching it to boundary conditions assumed to hold at the surface of the star. The only solution which has been well studied is that for a rotating spherical perfect conductor carrying a large magnetic dipole moment solved in flat spacetime.

The structure of the Einstein–Maxwell equations has been known for a long time but a major difficulty in obtaining specific solutions has involved the coupled structure of the eight partial differential equations in six unknowns. Standard devices which lead to a decoupling of the equations in a flat spacetime fail in a curved geometry. Maxwell's equations are

$$\nabla_\mu \mathbf{f}_{\nu\lambda} + \nabla_\nu \mathbf{f}_{\lambda\mu} + \nabla_\lambda \mathbf{f}_{\mu\nu} = 0 \tag{4.4}$$

$$\nabla_\mu \mathbf{f}_{\mu\nu} = 0 \tag{4.5}$$

where ∇_μ denotes covariant differentiation and $\mathbf{f}_{.\mu\nu}$ is the

Maxwell tensor

$$f_{\mu\nu} = \nabla_\mu A_\nu - \nabla_\nu A_\mu \tag{4.6}$$

A major computational advance has been achieved with the use of two-component Hertz potential formalism (see Nisbet and Cohen and Kegeles). For the Schwarzschild metric (non-rotating neutron star, see Appendix C) the electric and magnetic multipole field components are

$$E_1 = \frac{l(l+1)}{r^2}\exp(-ikt)R_{lk}(r)Y_{lm}(\theta,\phi),\quad B_1 = 0$$

$$E_2 = \frac{\exp(-ikt)}{Br}\frac{d}{dr}(R_{lk}(r))\frac{\partial}{\partial\theta}(Y_{lm}(\theta,\phi)),$$

$$B_2 = \frac{\exp(-ikt)}{Ar\sin\theta}R_{lk}(r)Y_{lm}(\theta,\phi)$$

$$E_3 = \frac{im\exp(-ikt)}{Br\sin\theta}\frac{d}{dr}(R_{lk}(r))Y_{lm}(\theta,\phi),$$

$$B_3 = \frac{ik\exp(-ikt)}{Ar}R_{lk}(r)\frac{\partial}{\partial\theta}(Y_{lm}(\theta,\phi)) \tag{4.7}$$

where the $R_{lk}(r)$ are the solutions of the wave equation

$$k^2R_{lk} + \frac{A}{B}\frac{d}{dr}\frac{A}{B}\left(\frac{dR_{lk}}{dr}\right) - \frac{l(l+1)}{r^2}\frac{A}{B}R_{lk} = 0 \tag{4.8}$$

and tend to spherical Bessel functions in the flat space limit

$$R_{lk}(r) \xrightarrow[r\to\infty]{} rj_l(kr) \tag{4.9}$$

while A and B are the metric functions

$$A^2(r) = B^{-2}(r) = 1 - \frac{2GM}{rc^2} \tag{4.10}$$

Equations (4.7) give the electric multipoles (except for $l = 0$ in which case $E_1 = 1/r^2$ independent of k) for the orthonormal coordinate frame

$$dx^0 = A dt,\ dx^1 = B dr,\ dx^2 = r d\theta,\ dx^3 = r \sin\theta\, d\theta \tag{4.11}$$

Similarly solutions may be obtained for the Kerr metric (rotating black hole)

$$\begin{aligned} ds^2 = &-(1-2Mr/\Sigma)dt^2 - \{4Lr \sin^2\theta/\Sigma\}dt d\phi + \\ &+ (\Sigma/\Delta)dr^2 + \Sigma d\theta^2 + (\sin^2\theta)\{(r^2 + L^2/M^2)^2 - \\ &- (\Delta^2 \sin^2\theta/M^2)\}/\Sigma\ d\phi^2 \end{aligned} \tag{4.12}$$

with $\Sigma = r^2 + L^2/M^2 \cos^2\theta$ and $\Delta = r^2 - 2Mr + L^2/M^2$ where L is the angular momentum of the star. For details the reader is referred to the literature (e.g. Cohen and Kegeles). The components of the Hertz field satisfy the wave equation

$$\begin{aligned} &\left\{\frac{(r^2+L^2/M^2)^2}{\Delta} - \frac{L^2}{M^2}\sin^2\theta\right\}\frac{\partial^2\psi}{dt^2} + \frac{4lr}{\Delta}\frac{\partial^2\psi}{\partial t\partial\phi} + \\ &+ \left(\frac{L^2}{M^2\Delta} - \frac{1}{\sin^2\theta}\right)\frac{\partial^2\psi}{\partial\phi^2} - \Delta\frac{\partial^2\psi}{\partial r^2} - \frac{1}{\sin\theta}\frac{\partial}{\partial\theta}\sin\theta\frac{\partial\psi}{\partial\theta} + \\ &+ 2\left\{\frac{L(r-M)}{M\Delta} + i\frac{\cos\theta}{\sin^2\theta}\right\}\frac{\partial\psi}{\partial\phi} + \\ &+ 2\left\{\frac{M(r^2-L^2/M^2)}{\Delta} - r - \frac{iL\cos\theta}{M}\right\}\frac{\partial\psi}{\partial t} + \frac{1}{\sin^2\theta}\psi = 0. \end{aligned} \tag{4.13}$$

The appropriate metric for a particular rotating neutron star model depends on the specific equation of state for the star (see §1.2, §6.2, and Appendix C).

The self-consistent coupling of the fields to the magnetospheric plasma has not yet been attempted, and we shall content ourselves by describing, at the end of the next section, how the induced electric fields can lead to the injection of charged particles from the 'atmosphere' of the star into the magnetosphere.

4.2. The neutron star 'atmosphere'

As we approach the surface of the star the density of matter in the magnetosphere increases extremely rapidly. We noted in chapter 1 that at temperature $T \sim 10^7$ K the enormous gravitational forces at the surface of a neutron star led to a scale height for atomic atmospheres of the order of 5 cm. We shall in this section discuss the region in which matter is still basically atomic in nature, i.e. densities $\lesssim 10^4$–10^5 g cm^{-3}. The properties of matter in this region are still dominated by the magnetic field of the star.

Consider a hydrogen atom in a uniform magnetic field $B = 10^{12}$ G. We shall use the field to define the z-axis. The interaction energy of the electron spin with the field is

$$E_{\text{electron}-B} \simeq -\mu_B \cdot \mathbf{B} \simeq 10 \text{ keV} \tag{4.14}$$

while the corresponding proton spin interaction energy is

$$E_{\text{proton}-B} \simeq -\mu_N \cdot \mathbf{B} \simeq 5 \text{ eV} \tag{4.15}$$

and the proton–electron Coulomb energy is on the scale

$$E_{\text{electron}-\text{proton}} \simeq 10 \text{ eV} \tag{4.16}$$

For comparison, at a temperature $T \simeq 10^6$ K the thermal energy per particle is ~ 86 eV. Clearly the dominant feature is the interaction of the electron spin with the magnetic field, and the other interactions can be treated in perturbation theory. At temperatures ~ 10^6–10^7 K all the electrons will be essentially 100 per cent polarized while the proton spins will be almost randomly orientated.

Classically, the motion of an electron in a uniform magnetic field is well-known; the electron simply spirals along the magnetic lines of force, its motion governed by the Lorentz force (Fig. 4.1). Quantum-mechanically we must solve the

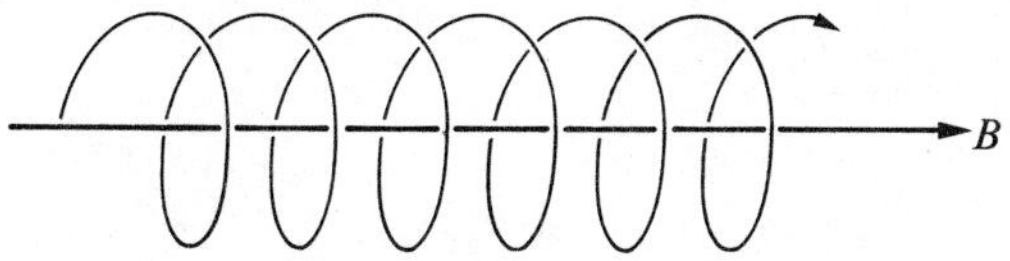

Fig. 4.1. Electron spiral orbit in a uniform magnetic field.

Schrödinger equation [the interaction energy scales of eqns (4.14)–(4.16) indicate that a nonrelativistic treatment is adequate.]

$$\frac{\left(\mathbf{p} - \frac{e}{c}\mathbf{A}\right)^2}{2m_e}\psi_n = E_n\psi_n \tag{4.17}$$

or

$$\left\{-\frac{\hbar^2}{2m_e}\nabla^2 + \frac{ie\hbar}{2m_e c}\mathbf{A}\cdot\nabla + \frac{e^2}{2m_e c^2}A^2\right\}\psi_n = E_n\psi_n \tag{4.18}$$

since

$$\nabla\cdot\mathbf{A} = 0 \tag{4.19}$$

for the Lorentz gauge in a static field. For a uniform **B** along the z-axis the vector potential is given by $\mathbf{B} = \nabla\wedge\mathbf{A}$, whence

$$\mathbf{A} = \tfrac{1}{2}B(-y, x, o) \tag{4.20}$$

so that the Schrödinger equation (4.18) becomes

$$-\frac{\hbar^2}{2m_e}\nabla^2\Psi_n(r) + \frac{eB}{4m_e c}\hbar i\left(y\frac{\partial}{\partial x}-x\frac{\partial}{\partial y}\right)\Psi_n(r) + \frac{e^2B^2}{8m_e c^2}(x^2+y^2)\Psi_n(r) =$$

$$= E_n\Psi_n(r) \tag{4.21}$$

If it were not for the second term, this would be separable in either Cartesian coordinates or cylindrical polar coordinates. In any case the z-component of the wavefunction is clearly a momentum-conserving plane wave $\sim \exp(ikz)$ contributing an amount $\hbar^2k^2/2m_e$ to the total energy. The second term on the left hand side of eqn (4.21) is simply proportional to the z-component of angular momentum which, since the problem has cylindrical symmetry, will be conserved, and to which we shall assign the quantum number m. This then yields an azimuthal function $e^{im\phi}$ contributing $-(e\hbar/2m_e c)Bm$ to the total energy. Finally there is a radial function satisfying the equation

$$-\frac{\hbar^2}{2m}\left\{\frac{1}{\rho}\frac{\partial}{\partial\rho}\frac{\partial}{\rho\partial\rho} - \frac{m^2}{\rho^2} + k^2\right\}R_{n_\rho}(\rho) - \frac{eBm\hbar}{2m_e c}R_{n_\rho}(\rho) + \frac{e^2B^2}{8m_e c^2}\rho^2R_{n_\rho}(\rho)$$

$$= E_{n_\rho}R_{n_\rho}(\rho) \tag{4.22}$$

i.e. a two-dimensional harmonic oscillator of oscillator energy

$$\hbar\omega = \mu_B B = \left(\frac{e\hbar}{2m_e c}\right)B \tag{4.23}$$

The oscillator eigenfunction is given in terms of the confluent hypergeometic functions

$$R \sim e^{-\rho^2/2b^2}(\rho/b)^{|m|}(\rho/b)^{|m|/2}F(n_\rho,|m|+1,(\rho/b)^2) \tag{4.24}$$

and contributes to the total energy an amount

$$E_{n_\rho} = 2(n_\rho + \tfrac{1}{2}|m| + \tfrac{1}{2})\hbar\omega \tag{4.25}$$

while the radial dimensions are given by the oscillator size parameter b

$$b = (2\hbar c/eB)^{\frac{1}{2}} \tag{4.26}$$

Thus the total energy can be written in the form

$$E_{n_\rho m}(k) = 2(n_\rho + \tfrac{1}{2}|m| - \tfrac{1}{2}m + \tfrac{1}{2})\hbar\omega + \hbar^2 k^2/2m_e. \tag{4.27}$$

In a field $B = 10^{12}$ G the oscillator energy $\hbar\omega \simeq 10$ keV and hence to excite anything other the z and azimuthal motion requires an energy input on this scale. Thus, to a very good approximation we need only consider the oscillatory ground state for which $n_\rho = 0$. Note that this state is infinitely degenerate for any integer quantum number $m \geqslant 0$. This degeneracy is associated with the freedom of choice of gauge for $\mathbf{A}$, which has the effect of moving the z-axis parallel to itself and hence changing the z-component of angular momentum. Physically, as m increases so does the kinetic energy, but this is exactly compensated by the increased interaction between the magnetic field and the orbital magnetic moment.

For motion centred on the z-axis the allowed spiral orbits have quantized radii

$$\rho_m = (2m+1)^{\frac{1}{2}} b \tag{4.28}$$

and in a field of 10^{12} G we have for the ground state orbit $\bar{\rho}_0 = b \simeq 10^{-10}$ cm, i.e. over an order of magnitude less than the radius of a terrestrial hydrogen atom.

So far we have discussed the motion of an isolated electron in a uniform magnetic field. We now turn our attention to the structure of an isolated atom in a huge magnetic field, and we shall consider first a hydrogen atom. From eqns (4.14) and (4.16) we see that we can treat the effect of the proton on the electron orbit in perturbation theory, treating the Coulomb potential e^2/r as a perturbative interaction. Also it is clear that the Coulomb potential is unlikely to excite the higher oscillatory states and only the z-component of the wavefunction will experience any detectable perturbation. If the higher oscillatory states are not excited, we may approximate the perturbing Coulomb potential as follows

$$\frac{e^2}{r} \equiv \frac{e^2}{\sqrt{(\rho^2+z^2)}} \simeq \frac{e^2}{\sqrt{(\rho_0^2+z^2)}} = \frac{e^2}{\sqrt{(b^2+z^2)}} \tag{4.29}$$

The approximation (4.29) is not as drastic as it might at first sight seem, since the radius of the electron spiral in a strong magnetic field is rather well defined $[\Delta\rho_0/\bar{\rho}_0 \lesssim 0.5]$. Thus the energy levels and eigenstates of the hydrogen atom in a huge magnetic field are approximately given by

$$\left[-\frac{\hbar^2}{2m_e}\frac{\partial^2}{\partial z^2} - \frac{e^2}{\sqrt{(b^2+z^2)}}\right]\Phi_{n_z}(z) = E_{n_z}\Phi_{n_z}(z). \tag{4.30}$$

The perturbation being itself cylindrically symmetric cannot change the z-component of angular momentum, thus the total wavefunction is of the form

$$\Psi(\mathbf{r}) \sim R_0(\rho)\exp(im\phi)\Phi_{n_z}(z) \tag{4.31}$$

and the total ground state perturbed energy is

$$E_{00n_z} = \hbar\omega + E_{n_z} \tag{4.32}$$

Equation (4.30) can be solved numerically but we can obtain considerable physical insight into the solution by studying its asymptotic forms. At large distances $z \gg b$ we have

$$\frac{e^2}{\sqrt{(b^2+z^2)}} \simeq \frac{e}{z}\left(1-\tfrac{1}{2}\frac{b^2}{z^2}+\ldots\right) \tag{4.33}$$

Thus asymptotically we have at large distances a one-dimensional Coulomb potential confining the electrons in the z-direction. At very short distances $z \ll b$ we have

$$\frac{e^2}{\sqrt{(b^2+z^2)}} \simeq \frac{e^2}{b}\left(1-\tfrac{1}{2}\frac{z^2}{b^2}+\ldots.\right) \tag{4.34}$$

i.e. a one-dimensional oscillator well of depth $e^2/b \simeq 1$ keV and an oscillator energy $h\omega_z \simeq e\hbar/(m_e b^3)^{\frac{1}{2}} \simeq 300$ keV. Thus, the oscillator zero point energy is much greater than the depth of the well; hence any binding of the electron to the proton is provided by the long range Coulomb tail. The complete perturbative potential is illustrated in Fig. 4.2. We note that the long range asymptotic form of (4.30)

$$\left(-\frac{\hbar^2}{2m_e}\frac{\partial^2}{\partial z^2} - \frac{e^2}{z}\right)\phi_{n_z}(z) = E_{n_z}\phi_{n_z}(z) \tag{4.35}$$

is identical with the reduced radial equation for the ground

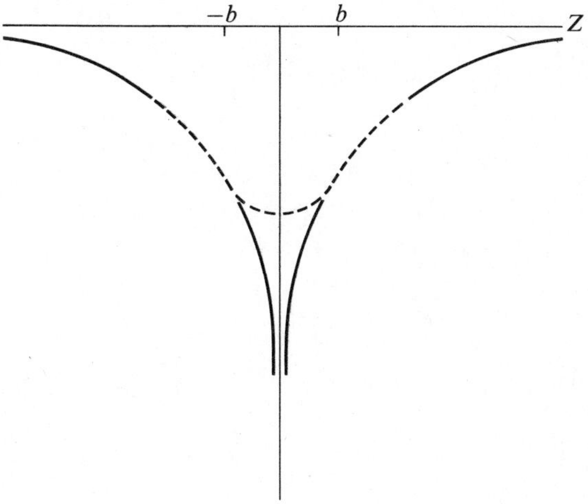

Fig. 4.2. The asymptotic ($\sim 1/z$) Coulomb potential is modified at short distances ($\sim b$) becoming an oscillator potential ($\sim z^2$) at very short distances ($\ll b$).

state (s-state) of the usual spherical hydrogen atom

$$\left(-\frac{\hbar}{2m_e}\frac{\partial^2}{\partial r^2} - \frac{e^2}{r}\right)\phi_0(r) = E_0\phi_0(r) \tag{4.36}$$

The usual ground state $\phi_0(r)$ has a scale size given by the Bohr radius $a_0 \gg b$ so that the asymptotic long range form (4.35) is appropriate. Thus the solutions of (4.35) will accurately be the one-dimensional analogues of the normal hydrogen radial s-states, they will have z-dimensions of order a_0 and the binding energies will be $\lesssim 13.6$ eV. Note that as B increases b decreases, and this approximation gets better and better. These statements, however, do not apply to the *ground state* of eqn (4.35), because in one dimension we have $-\infty \leqslant z \leqslant +\infty$ while in three dimensions we have $0 \leqslant r \leqslant \infty$. Thus the ground state of eqn (4.36) which has $\phi_0(r=0) = 0$ corresponds to the first excited state of eqn (4.35) which has a single node at $z = 0$ [see Fig. 4.3]. The ground state of eqn (4.35) with no node in the z-direction corresponds to an approximately cylindrical electronic distribution of length l and radius b, thus the energy is approximately

$$\varepsilon_0 \sim \hbar^2/m_e l^2 - \frac{e^2}{l}\ln\left(\frac{l}{b}\right) \tag{4.37}$$

and minimizing with respect to l yields

$$\frac{2a_0}{l} \simeq \ln\left(\frac{l}{b}\right) - 1 \tag{4.38}$$

or very roughly [for large B, $b \ll l$]

$$l \simeq \frac{2a_0}{\ln a_0/b} \simeq \frac{a_0}{2}. \tag{4.39}$$

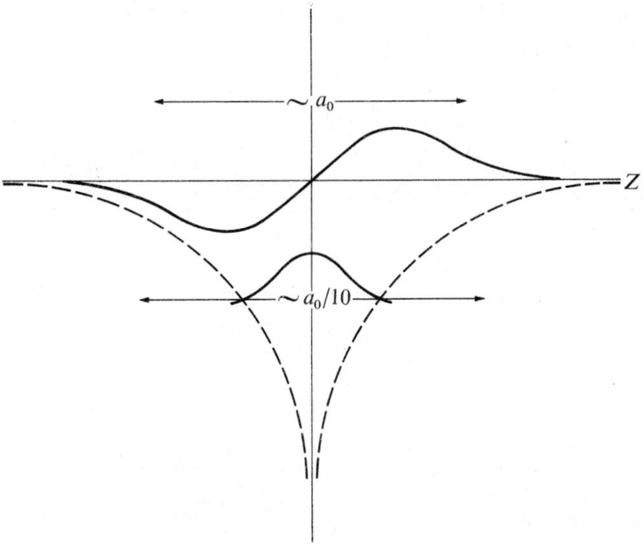

Fig. 4.3. The first two states (unnormalized) of the one-dimensional Coulomb potential. In three dimensions the radius variable is in the interval $0 \leqslant r \leqslant +\infty$, and only the right-hand side of the above diagram has any meaning. In this case the lowest state has no meaning and the first excited state of the one-dimensional case can be associated with the ground state of the three-dimensional case.

Substituting (4.39) into (4.37) we find that the ground state has a binding energy

$$\varepsilon_0 \sim -\frac{\hbar^2}{4ma_0^2}\left[\ln\left(\frac{a_0}{b}\right)\right]^2 \sim -100 \text{ eV}. \tag{4.40}$$

In equation (4.27) we saw that there was an infinite series of degenerate states corresponding to different z components of angular momentum $m\hbar$, $m = 0, 1, 2 \ldots. + \infty$, and that the radius of maximum probability for such orbits is $\rho_m = (2m+1)^{\frac{1}{2}}b$. So the second effect of the Coulomb potential is to split the degeneracy of such orbits by an amount $\sim l^2/a_0\ (b^2/a_0^2) \sim 0.1$ eV.

Thus, there are three modes of excitation of our distorted hydrogen atom corresponding to the quantum numbers n_ρ, m, n_z. The ground state ($n_\rho = m = n_z = 0$) is bound by an energy ~ 100 eV, and the lowest excited states ($n_\rho = n_z = 0$, $m = 0, 1,$

...) or Landau orbitals have excitation energies of the order of eV. Next come the z-component excitations ($n_\rho = 0$, $n_z = 1, 2, \ldots$, $m = 0, 1 \ldots$) which are at an excitation energy of ~ 90 eV and finally there are the excited oscillator modes ($n_\rho = 1$, $n_z = 0, 1, 2 \ldots$, $m = 0, 1, 2 \ldots$) which begin at excitation energies ~ 10 keV. The full spectrum is schematically illustrated in Fig. 4.4.

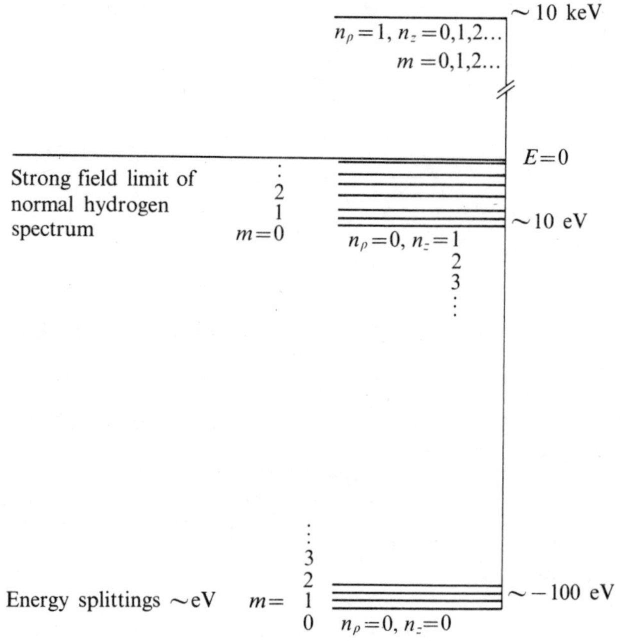

Fig. 4.4. Spectrum of hydrogen atom in a field ~ 10^{12} G.

Turning now to multielectron atoms we see that we have lost the systematics of the periodic table of elements which depends on the special degeneracies associated with the spherical symmetry of the Coulomb potential. At least for light atoms, successive electrons simply populate successive Landau orbitals, all spins being antiparallel to the magnetic field. Thus the length of the atoms is essentially independent of the charge number Z while the radius varies roughly as $(2Z+1)^{\frac{1}{2}}$. Hartree–Fock calculations suggest that the ionization energy of the light atoms varies as ln Z and shows none of the sharp peaks and valleys associated with a spherical shell structure.

This successive filling of Landau orbitals breaks down when it becomes energetically more favourable for the last electron to populate the $n_z = 1$, $m = 0$ orbital rather than the Landau orbital $n_z = 0$, $m = Z$. This occurs when the electron probability density is closer to the nucleus for the $n_z = 1$, $m = 0$ orbital than for the $n_z = 0$, $m = Z$ orbital. We have seen that the radii of the Landau orbitals are $\sim (2Z+1)^{\frac{1}{2}}b$ while, remembering the similarity between the z-component of the Schrödinger equation of the distorted atoms and the reduced spherical hydrogenic equations, we have that for $n_z = 1$ the appropriate length scale is a_0/Z. Thus a useful parameter is

$$\eta = \frac{a_0}{Z(2Z+1)^{\frac{1}{2}}b} \simeq 10^{-5}\left(\frac{B}{Z^3}\right)^{\frac{1}{2}} \tag{4.41}$$

where B is measured in Gauss. For $\eta \gg 1$ we have the filling of Landau orbitals and completely cylindrical atoms. As η decreases the centre of the cylinders start to fill, and for $\eta \ll 1$ we have essentially spherical atoms. In the intermediate region $\eta \sim 1$ we have a spherical core which shields the outer electrons (i.e. reduces the effective Z), and these valence electrons again populate Landau orbitals giving rise once more to a basically cylindrical shape, see Fig. 4.5. The most massive nucleus that we expect to find in the low density region of the neutron star is ^{56}Fe with $Z = 26$. The next most abundant element we might expect to be ^{4}He with $Z = 2$. Hence for magnetic fields $\gtrsim 10^{12}$ G we are in a region where all the atoms may be expected to be highly distorted and essentially cylindrical in shape with the axis of the cylinders aligned with the magnetic field.

The chemistry of terrestrial elements is dictated by the fact that normally atoms are nearly spherical. The interaction between neutral atoms is due to the small deviations from sphericity, and atomic bonds are due to valency and polarizability. Due to the smallness of these effects atomic bonds are relatively weak, and the binding energy of a molecule is

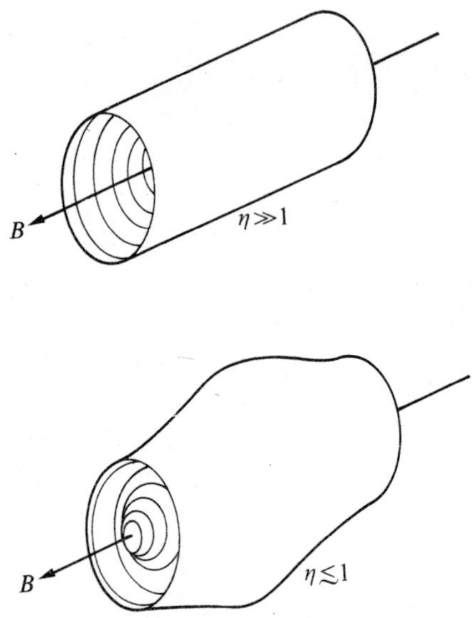

Fig. 4.5. Distorted shapes of atoms in huge magnetic fields. For $\eta \gg 1$ [eqn (4.37)] the atoms are essentially cylindrical. As η decreases the atoms develop a spherical core and finally for $\eta \ll 1$ the atoms are completely spherical.

usually considerably less than the binding energy of the atoms of which it is composed. When two atoms approach each other their interaction is determined by the state of the least bound (valence) electrons. These are the electrons which are disturbed by the presence of the other atom and which are shared in the covalently bound molecule. Consider the normal hydrogen molecule in which the spins of the two electrons are parallel; we may for instance imagine them polarized by a weak magnetic field. When the two atoms are brought together the electrons are initially in $1s$ states, but since the spins are parallel the Pauli exclusion principle requires that one of the electrons be excited to the $2s$, $2p$ ... state. The energy of this excitation is greater than the bonding energy so that the triplet state of the normal hydrogen molecule is unbound. Now let us consider the interaction between two magnetically distorted hydrogen atoms in a huge magnetic field. The cylindrical shapes give rise to enormous electric quadrupole moments and the quadrupole–quadrupole interaction is extremely strong without the need for any further polarization. Two distorted hydrogen atoms, initially in their ground states, approaching each other would have their electrons in $m = 0$ Landau orbitals and the highly directional quadrupole–quadrupole interaction would favour bonding along the axis of the cylinders. Because of the Pauli principle, one of the electrons would have to be excited to the $m = 1$ Landau orbital, but this excitation energy is only of the order ~ 1 eV compared with the electron binding energy ~ 100 eV; the resulting molecule is extremely tightly bound. A series of hydrogen atoms would be expected to slot together rather like a tent pole to form a polymer chain, see Fig. 4.6.

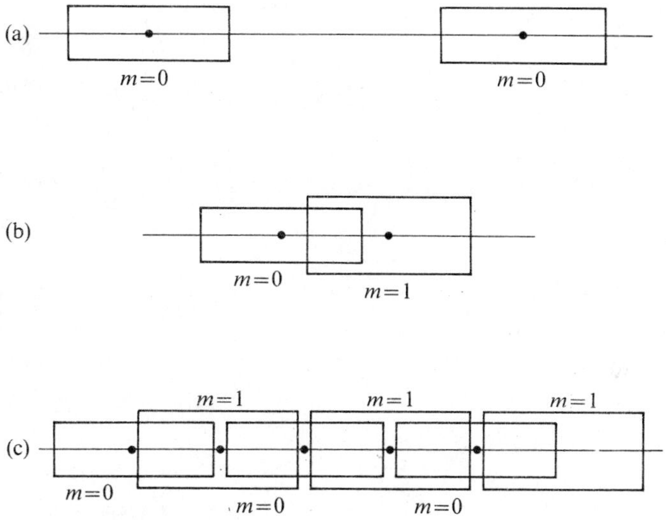

Fig. 4.6. (a) Two isolated hydrogen atoms distorted by a huge magnetic field. In both atoms the lowest Landau orbital is occupied. (b) In the hydrogen molecule the Pauli principle requires one electron to be in the first excited Landau orbital. (c) A polymeric chain of hydrogen atoms.

We found the binding energy of an isolated hydrogen atom to be $\sim e^2/a_0\ [\ln\eta]^2$ [see eqn (4.40)]. The binding energy per atom in an infinitely long hydrogen polymeric chain is $(e^2/a_0)\eta^{4/5}$ and thus for $\eta \gg 1$ the chain is more stable than an isolated atom. This situation persists for multielectron atoms provided $\eta \gg 1$ and we find that the total binding energy of an isolated atom is $\sim (Z^3e^2/a_0)[\ln\eta]^2$ while the binding energy per atom in the polymeric chain is $(Z^3e^2/a_0)\eta^{4/5}$. For heavier atoms like iron η is not much greater than 1 and the atoms have a spherical core enclosed in a cylindrical sheath of 'valence' electrons (see Fig. 4.5). It is only the 'valence' electrons which contribute to the bonding and thus the polymeric bonding is much reduced compared with the $\eta \gg 1$ case. For ^{56}Fe we have 18 valence electrons and the binding energy per atom in an infinitely long polymer chain is $E_a \sim 10$ keV. The resulting structure of the polymer is schematically illustrated in Fig. 4.7.

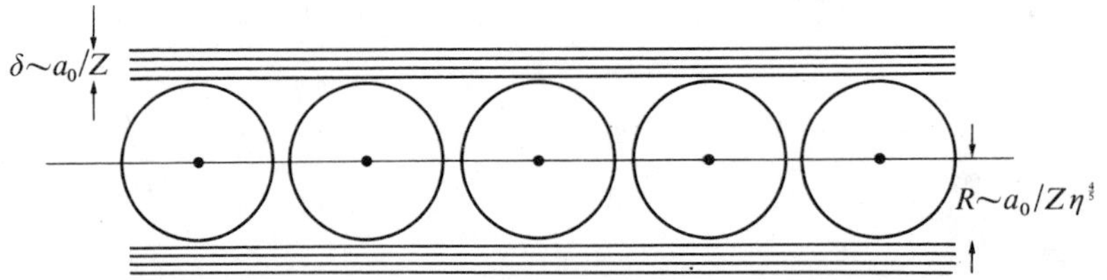

Fig. 4.7. In heavy atoms, $\eta \lesssim 1$, most electrons form a spherical core of radius R; the valence electrons form a cylindrical sheath around the polymer.

A polymeric chain will attract a neighbouring one even in the absence of Van der Waals forces simply because of the nature of the quadrupole–quadrupole force. The equilibrium configuration will then be a close-packed tetragonal space lattice, see Fig. 4.8. The Young's modulus for such a polymeric solid will be $Y \sim E_a/lb^2$ from which we obtain for iron $Y \sim 10^{19}$ dyne cm^{-2} or more than a million times greater than terrestrial steel.

The size of the atoms is $\sim lb^2$, and hence in the close-packed configuration the matter is $\sim 10^4$ times the density of a terrestrial solid. Since the polymeric bonding is ~ 10 keV we see that the solid is stable up to temperatures $\sim 10^8$–10^9 K, which are expected to be considerably in excess of the surface temperature of all but the very newest neutron stars. Thus, we do not expect a gaseous atmosphere on a neutron star, but rather that at even almost zero vapour pressure the surface material will condense to form a magnetic

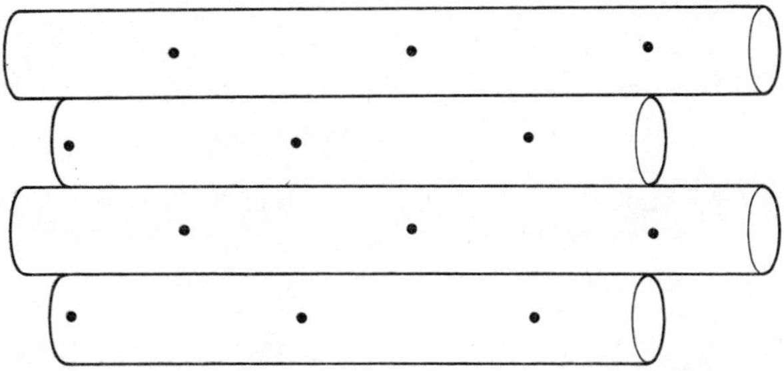

Fig. 4.8. The polymer space lattice.

polymeric solid and that the density will rise from zero to $\sim 10^4$ g cm^{-3} within a Bohr radius. There will then be a depth of $\sim$ 1 cm polymeric solid in which the density rises 'slowly' to $\sim 10^5$ g cm^{-3}. Below this atmosphere there is a crust at densities $\gtrsim 10^7$ g cm^{-3} where the structure is governed by the Coulomb forces between the bare nuclei and in which the electron density has increased to the point where electron capture by the nuclei has become significant so that the nuclear species becomes a rapidly varying function of density. This crust will be discussed in more detail in the next chapter.

The polymeric solid behaves like a one-dimensional metal being an excellent conductor parallel to the magnetic field and an insulator perpendicular to it. In the case $\eta \sim 1$ only the 'valence' electrons are conducting. The electron work function for the solid is expected to be typically $\omega \sim 100$ eV and only surface electric fields parallel to the magnetic field are likely to be effective in inducing field emission leading to the injection of electrons into the magnetosphere. A conducting sphere of radius R having a magnetic dipole field B and rotating with frequency ω in a vacuum generates a parallel surface field E_s^0 given by

$$E_s^0 \sim \frac{\omega R}{c} B; \tag{4.42}$$

this corresponds to $\sim 10^{12}$ V cm^{-1} for the Crab pulsar. Depending on the magnetosphere the actual surface electric field may be much less than this, i.e.

$$E_s \sim \varepsilon E_s^0 \tag{4.43}$$

where $\varepsilon = 1$ for no magnetosphere and $\varepsilon = 0$ for an exactly corotating magnetosphere. The field emission electron current is

$$j \simeq 1000\ E_s^2 \exp\left\{-\frac{2W\times10^{14}}{\varepsilon\omega B}\right\} (\mathrm{A\ cm^{-2}}) \tag{4.44}$$

where W is in keV, E_s in V $\mathrm{cm^{-1}}$, ω in $\mathrm{s^{-1}}$ and B in G. For the Crab pulsar most models yield $j \sim 10^6$ A $\mathrm{cm^{-2}}$. Note that the field-induced current falls rapidly with the age of the pulsar, i.e. decreasing ω.

4.3. Pulses

There are a number of models for the origin of pulsar signals, but no consensus view has yet been reached. Any model for the process must explain why all pulsars are observed at radio frequencies, but only the Crab pulsar is observed in the optical and x-ray regions as well.[1] It must also explain the limited range of pulsar periods from milliseconds to seconds.

We have seen in the previous section that charges may be driven out from the star along the magnetic field lines, which are curved. These accelerated charges will radiate. The radiation will be focussed in a cone emanating from the magnetic poles. As the star rotates an observer might see a lighthouse-like pulsed signal (Fig. 4.9). While this is probably a correct picture of the primary source of the pulses, in this simple form it fails to meet the criteria we have laid out above. Let us examine briefly how the model might be refined to account for the observations.

We note that the magnetic field lines which emanate from the magnetic polar regions and reach the velocity of light cylinder are open lines and that they must be twisted (see Fig. 4.10) if a particle tied to them is not to exceed the velocity of light. The radius of curvature ρ of the magnetic field at a radius r is

---

[1] At the time of writing the first observation of optical pulses from the Vela pulsar have just been reported and there are preliminary indications that a number of the faster pulsars may pulse at other than radio frequencies also.

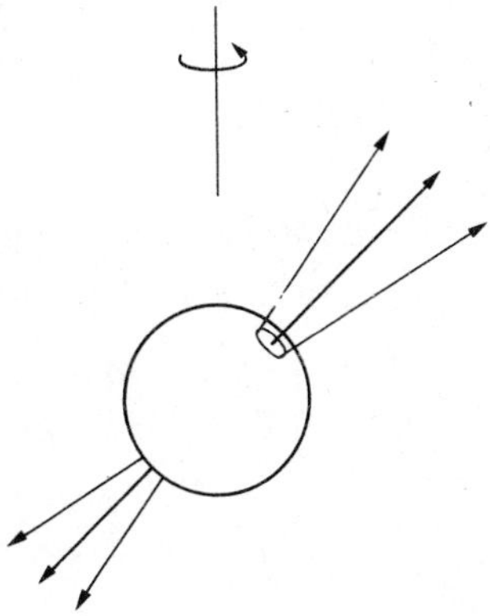

Fig. 4.9. Radiation from the polar regions of a rotating neutron star will produce a lighthouse-like effect.

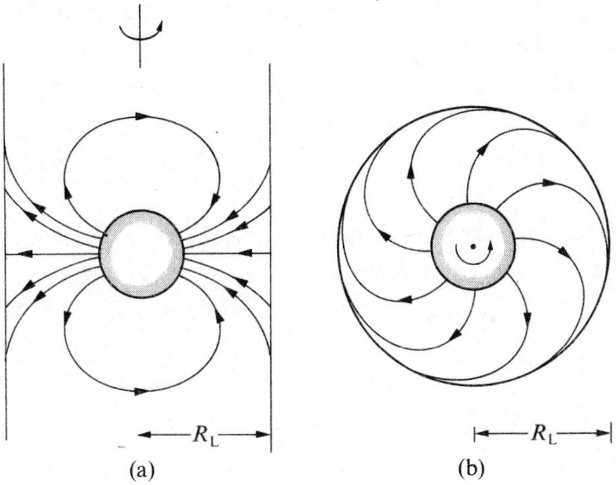

Fig. 4.10. The magnetic field lines of a rotating neutron star. For simplicity we have illustrated the case of a star with its magnetic axis normal to its rotation axis. (a) in the plane of the magnetic axis and the rotation axis we see that field lines emanating from the magnetic poles reach the velocity of light cylinder. (b) In a plane normal to the axis of rotation we see that the magnetic field lines curve onto the velocity of light cylinder.

$$\rho \sim (rc/\omega)^{\frac{1}{2}} \tag{4.45}$$

and the radius of the velocity of light cylinder R_L is

$$R_L = c/\omega \tag{4.46}$$

A relativistic particle of charge e and rest mass m_0 with energy $E \gg m_0c^2$ moving along an orbit with radius of curvature ρ will radiate the power spectrum

$$I(\nu) \simeq \frac{1}{2\pi}\frac{e^2}{c}\left(\frac{c}{\rho}\right)\frac{E}{m_0c^2}(\nu/\nu_0)^{1/3} \qquad \nu < \nu_0 \tag{4.47}$$

where the limiting frequency is

$$\nu_0 = \frac{1}{2\pi}\frac{c}{\rho}\left(\frac{E}{m_0c^2}\right)^3. \tag{4.48}$$

Primary charges leaving the polar regions will be accelerated along the magnetic field lines and radiate gamma rays tangentially. These gamma rays, if they are energetic enough, will give rise to electron–positron pair production. These secondary charges will be accelerated in opposite directions and will in turn give rise to more gamma rays which will lead to more pair production and thus a cascade will be established, Fig. 4.11.

Let us discuss what might happen if the primary charges are electrons. Their low mass means that they will quickly be

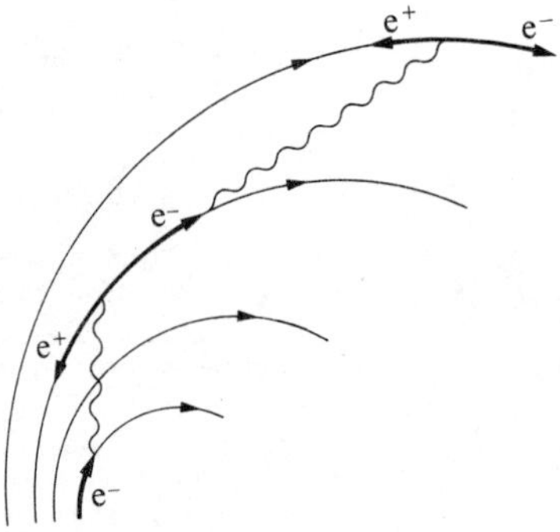

Fig. 4.11. A highly relativistic primary electron following the curved magnetic field line radiates a high energy γ-ray which subsequently produces an electron-positron pair and so on until a cascade is established.

accelerated to relativistic velocities, and their maximum energy is likely to be radiation reaction limited, i.e. they will reach an energy where the rate at which they are losing energy (eqn (4.47)) is equal to the rate at which they are gaining energy from the electric field (eqn (4.42)). The secondary pair production leads to the formation of charged sheets moving out along the magnetic field lines. The high frequency components of the resulting radiation field will suffer attenuation due to synchrotron self absorption and only the radio frequencies will survive. This mechanism for producing radio frequency pulses requires that the primary electrons produce γ-rays of sufficient energy to yield pair production. This means that the limiting frequency ν_0 of eqn (4.48) must be such that

$$h\nu_0 > 2m_e c^2. \tag{4.49}$$

Combining (4.49) with (4.45) and (4.48), we see that this implies that the rotational period should be

$$\tau < \frac{h}{2m_e c^2}\left(\frac{E}{m_e c^2}\right)^3 \tag{4.50}$$

In the radiation reaction limit for the energy E this implies $\tau \lesssim 5$ s.

Now let us consider what happens if the primary charges are protons. Because of their much greater mass they will initially be accelerated much more slowly. Indeed, it is likely that their energy will never reach the radiation reaction limit. The pair cascade will thus be established at a much greater distance from the surface of the star. Here the magnetic field strength is much weaker and the synchrotron self absorption peak will move to lower frequencies, i.e. into the radio frequency region, leaving the optical and x-ray pulses to escape. The critical rotational frequency for proton-induced pair production will be some $(2000)^{2/3}$ times greater than that for electron-induced pair production and hence an upper limit for the rotational period would be ~ 0.05 s.

Thus, we have a model in which only neutron stars rotating with a period less than ~ 0.05 s would pulse at the optical and x-ray frequencies. All others would pulse only at the radio frequencies and the whole process would automatically switch itself off for rotational periods in excess of ~ 5 s.

The lower limit on the rotational period is clearly related to the maximum rotational frequency consistent with stellar equilibrium, i.e.

$$R\omega^2 \lesssim \frac{GM}{R^2} \tag{4.51}$$

or

$$\omega \lesssim \surd(G\rho) \tag{4.52}$$

which leads to periods $\tau \gtrsim 0.001$ s (see Table 1.1).

Suggested references

LAMB, F.K., PETHICK, C.J., and PINES, D. (1973). *Astrophys. J.* **184**, 271. Studies the accretion of matter onto compact rotating magnetic stars.

TER HAAR, D. (1972). *Phys. Rep.* **3**, 57. Reviews some aspects of the neutron star magnetosphere.

STURROCK, P.A. (1971). *Astrophys. J.* **164**, 529. An interesting pulsar model.

NISBET, A. (1955). *Proc. Roy. Soc.* **A231**, 250. COHEN, J.M. and KEGELES, L.S. (1974). *Phys. Rev.* **D10**, 1070. Electromagnetism in curved space geometries.

CHEN, H.H., RUDERMAN, M.A., and SUTHERLAND, P.G. (1974). *Astrophys. J.* **191**, 473. Discussed the properties of matter in superstrong magnetic fields.

GOLDREICH, P. and JULIAN, H.W. (1969). *Astrophys. J.* **157**, 869. The classic paper on the extraction of ions from the surface of a rotating neutron star.

5

NEUTRON STAR STRUCTURE

5.1. Introduction

IN this chapter we shall outline the results of calculations to determine the structure of neutron stars and of attempts to derive equations of state which will allow calculations of mass–radii relationships. The procedure is to consider a region of neutron star matter at a given density, attempt to calculate the internal energy at that density, and minimize the internal energy with respect to various configurations of the components of the matter in order to obtain the equilibrium configuration. Clearly this approach is to some extent suspect because in general only a few configurations are tested and there may well be phases not considered by the calculator which have a lower internal energy. Such phase transitions will be missed in these calculations. It is likely, however, that such phases represent very small proportional changes in the internal energy and hence will not seriously affect the equation of state – they may, of course, seriously change other physical properties of the matter, e.g. conductivity, etc.

Once we have calculated the equilibrium internal energy E as a function of density ρ we proceed to calculate the pressure on the assumption that the product of the entropy S and the temperature T is negligible compared with the internal energy. This is basically the assumption that all the Fermi energies are very much greater than the thermal energies kT. In this case the Helmholtz free energy F is approximately equal to the internal energy,

$$F(\rho) = E - ST \simeq E(\rho) \tag{5.1}$$

and the pressure versus density equation of state then follows from

$$P(\rho) = -\left(\frac{\partial F}{\partial V}\right)_T \simeq \rho^2\left(\frac{\partial \varepsilon}{\partial \rho}\right)_T . \tag{5.2}$$

In detail, of course, we will have several components in the system, electrons, protons, neutrons, etc., and we will obtain the total pressure as a sum of partial pressures (some of which may be negative) for each of the components.

For purposes of calculating gravitational effects we need to know the effective contributing mass density and this is defined to be

$$\rho_{mass} = \varepsilon/c^2. \tag{5.3}$$

The surface of the star is then given by the locus of points at which the pressure defined by eqn (5.2) and the equation for stellar equilibrium (1.41) is zero.

For densities less than $\sim 3 \times 10^{14}$ g cm^{-3} we believe we know the forces and the species of particles present. This is the realm of matter at subnuclear densities. Most of the interest centres on possible many-body effects which can produce configurations of matter which cannot be realized in the small droplets of matter which comprise terrestrial atomic nuclei. At densities greater than $\sim 3 \times 10^{14}$ g cm^{-3} the problem becomes much more complicated; not only do the many-body problems become more complex, but even the species of particles present are not certain, nor are the relative abundances. As the density increases, and the relative abundance of more exotic baryons increases, the uncertainty in the interparticle interactions also increases and the predictions of the calculations must be considered more and more speculative.

In the appendices we have gathered together a brief survey of some of the calculational techniques employed and in this chapter we shall only quote sufficient detail necessary for an understanding of the predictions.

5.2. The nuclear crust

We know that terrestrial nuclei have a net binding energy, and hence at low densities we expect the nucleons to condense out to form nuclei. This is simply the logical limit of surface clustering studied in terrestrial nuclei except that there the mass numbers of the nuclei are small and hence the clustering is extremely limited —usually only alpha particle clustering is considered. These nuclei will be immersed in an electron gas whose density is such that the whole system is electrically neutral. If the density of matter ρ is much lower than the saturation density ρ_0 (2.8×10^{14} g cm^{-3}), then the mean spacing between the nuclei R_L will be much larger than the radii R_0 of the nuclei

$$R_L \simeq \left(\frac{\rho_0}{\rho}\right)^{1/3} R_0. \tag{5.4}$$

For surface densities of neutron stars $\rho \simeq 10^7$ g cm^{-3}, $R_L \simeq 300\ R_0$. Thus the nuclei interact with each other only through their screened Coulomb fields. This means that we expect the radii and neutron-to-proton ratios of the nuclei to be similar to those of terrestrial nuclei. Since the most stable terrestrial nucleus is ^{56}Fe, we expect this to be the most common species in the surface of the neutron star.

The concentration of charge into the nuclei gives rise to a screened Coulomb interaction between the nuclei. In seeking to minimize this interaction energy the nuclei form a space lattice.

Thus there are three contributions to the energy in this phase: there is the energy of the nuclei, the electron gas energy, and the lattice energy. As the density increases, the electron gas energy rises, and so does the lattice energy. Thus, we may expect a lower energy configuration in which some electrons are absorbed by the nuclei. This lowers the electron gas density and hence its energy, reduces the charge on the nuclei, and hence reduces the lattice energy. This is at the expense of nuclear stability, because, as the charge on the nuclei decreases without any change in the mass number, the neutron excess increases away from the line of beta stability. As the density increases, we would expect this process to continue until the nuclei become unstable against neutron emission. We now have a new phase containing a neutron gas and this we shall discuss more fully in the next section.

The electron gas energy density is

$$E_{\rm e} = \frac{1}{\pi^2\hbar^3}\int_0^{p_{\rm F}} (p^2c^2+m_{\rm e}^2c^4)^{\frac{1}{2}}p^2\,{\rm d}p$$

$$= \frac{c}{\pi^2\hbar^3}\Big[\tfrac{1}{4}p_{\rm F}(p_{\rm F}^2+m_{\rm e}^2c^2)^{3/2} - \tfrac{1}{4}m_{\rm e}^2c^2\Big\{\tfrac{1}{2}p_{\rm F}(p_{\rm F}^2+m_{\rm e}^2c^2)^{\frac{1}{2}} +$$

$$+ \tfrac{1}{2}m_{\rm e}^2c^2\ln\left(\frac{p_{\rm F}}{m_{\rm e}c} + \left(\frac{p_{\rm F}^2}{m_{\rm e}^2c^2}+1\right)^{\frac{1}{2}}\right)\Big\}\Big]$$

$$= W_{\rm e}n_{\rm e} \tag{5.5}$$

where the electron number density $n_{\rm e}$ is given by

$$n_{\rm e} = \frac{8\pi}{3h^3}p_{\rm F}^3 \tag{5.6}$$

and $W_{\rm e}$ is the energy per electron.

Equation (5.5) has the expected low density limit

$$W_{\rm e} \xrightarrow[n_{\rm e}\to 0]{} m_{\rm e}c^2 \tag{5.7}$$

and the familiar high density limit

$$W_{\rm e} \xrightarrow[n_{\rm e}\to\infty]{} \tfrac{3}{4}p_{\rm F}c \tag{5.8}$$

corresponding to the extreme relativistic situation. The limiting expression (5.8) is sufficiently accurate for most of our needs especially for densities in excess of $\sim 10^9$ g cm^{-3}.

The lattice energy is most simply calculated using the Wigner–Seitz approximation. Divide the lattice into unit cells with a single nucleus at the centre of each cell. Since each cell is electrically neutral, we can, to a first approximation, neglect the interaction between cells. The Coulomb energy of each cell may be estimated by replacing it with an equivalent sphere of radius r_L such that

$$4/3\ \pi r_L^{\ 3} n_N = 1 \tag{5.9}$$

where n_N is the number density of nuclei, and assuming that the Z electrons uniformly fill the cell. Then the lattice energy density is the total Coulomb energy of the cell less the self-Coulomb energy of the nucleus, i.e.

$$W_L = -\frac{9}{10}\frac{Z^2e^2}{r_L}\left(1 - \frac{5}{9}\frac{\langle r^2\rangle}{r_L^{\ 2}}\right) \tag{5.10}$$

where $\langle r^2\rangle$ is the mean square radius of the nuclear charge distribution. For a uniform charge distribution throughout a spherical nucleus of radius r_N we have

$$\langle r^2\rangle = \frac{3}{5}r_N^{\ 2}. \tag{5.11}$$

Finally we have the energy of the nuclei. For extremely low densities $\rho \lesssim 10^8$ g cm^{-3} we expect the nuclei to be ^{56}Fe and its neighbours not far from the line of beta stability for isolated nuclei, and the various semi-empirical mass formulae may be used to calculate the nuclear energy. As the density increases, the neutron excess in the nuclei rapidly increases, and the semi-empirical mass formulae become unreliable. There are a number of techniques for calculating nuclear energies and some of these are reviewed in Appendix A; here we simply note that in general these calculations yield a nuclear energy which is a function W_N of the mass number A, charge number Z, and nucleon density distribution $\rho_N(r)$,[1] which allows for the effects of surface thickness and possible distortions.

The total energy density is then

[1] Note that we include in the nuclear energy the self-Coulomb energy of the nucleus and the rest mass energy of the nucleons.

$$E_{\text{tot}}(A,Z,\rho_N,r_L,n_e n_N;\rho) = W_e n_e + W_L(Z,r_L,\rho_N)n_N + W_N(A,Z,\rho_N)n_N \tag{5.12}$$

where the nuclear density dependence in W_L comes implicitly from the term $\langle r^2 \rangle$ in eqn (5.10). For each density ρ the total energy density is minimized with respect to its independent arguments in order to establish the equilibrium components and configurations at that density. The total list of arguments in (5.12) is overcomplete due to various overall constraints, i.e. charge neutrality, which implies

$$n_e = Zn_N \tag{5.13}$$

and the mass density relationship

$$\rho = m_e n_e + (Zm_p + (A-Z)m_N)n_N \tag{5.14}$$

Thus, at a fixed density ρ, there are four independent argu= ments in E_{tot}; A and Z which determine the nuclear species, ρ_N which determines the nuclear structure, and r_L which determines the lattice structure.

5.3. The neutron drip region

The minimum energy required to add a neutron to a nucleus, i.e. the chemical potential of the neutrons in the nuclei is

$$\mu^{(N)} = \frac{\partial}{\partial A}(W_N + W_L)\Big|_{Z,\rho_N,r_L} \tag{5.15}$$

and similarly the proton chemical potential is

$$\mu_p^{(N)} = \frac{\partial}{\partial Z}(W_N + W_L)\Big|_{A-Z,\rho_N,r_L} \tag{5.16}$$

while the condition for beta stability is

$$\mu_e = (\varepsilon_F^e + m_e c^2) = \mu_n^{(N)} - \mu_p^{(N)}. \tag{5.17}$$

As long as $\mu_n^{(N)}$ and $\mu_p^{(N)}$ remain less than the free neutron and proton masses respectively, i.e. the Fermi energies remain negative, then there are no nucleons outside the nuclei. However, as the density increases so do the chemical potentials. Because of the Coulomb energy, $\mu_n^{(N)}$ rises faster than $\mu_p^{(N)}$, i.e. there is an increasing neutron excess. When the chemical potential of a neutron in the nucleus reaches the mass of the free neutron,[1] then neutrons will start to drip out of the nuclei. In this new regime we have nuclei immersed in a completely relativistic electron gas and a neutron gas. In equilibrium the chemical potential of a neutron in the gas $\mu_n^{(G)}$ must equal the chemical potential of a neutron in the nuclei $\mu_n^{(N)}$.

The calculation of the equilibrium configuration then proceeds as for the crust region described in §5.2 except that we now add a term $W_n(\rho_n^{(G)})n_n$ representing the energy density of the neutron gas. There is one other important modification, and that is the influence of the neutron gas on the energy of the nuclei. The neutron gas acts on the nuclei like a detergent on an oil drop in that it reduces the surface energy of the nuclei, and it also disturbs the neutron–proton density distributions in the nucleus. So now W_N is no longer simply a function of A, Z, and ρ_N but is a function of $\rho_n^{(N)}$, $\rho_p^{(N)}$, and $\rho_n^{(G)}$. The calculation of the surface energy can be carried out using the methods described in Appendix A. The onset of the neutron drip region occurs at $\sim 2 \times 10^{11}$ g cm^{-3}, and hence the nuclei still occupy only $\sim 10^{-3}$ of the volume. Interactions, other than the Coulomb forces, between the nuclei are still negligible, and the extent of the nuclei is still fairly well defined. Certainly the charge distribution $\rho_p^{(N)}$ is well defined and is related to the charge number Z by

---

[1] In practice slightly less than the mass of the free neutron since a neutron in the space around the nuclei acquires an effective mass which is slightly less than the free mass because of its interaction with the nuclei in the lattice.

$$Z = \int_{\text{u.c.}} \rho_p^{(N)}(\mathbf{r})\, d\mathbf{r} \tag{5.18}$$

where the integral extends over a unit cell. The root mean square radius of the charge distribution required for the calculation of the lattice energy

$$\langle r^2 \rangle = \int_{\text{u.c.}} \rho_p^{(N)}(\mathbf{r})\, r^2 d\mathbf{r} \tag{5.19}$$

is also well defined. The neutron distribution is not quite so clearly defined since the neutrons in the nucleus $\rho_n^{(N)}$ merge smoothly into the neutrons in the gas $\rho_n^{(G)}$. However, while the nuclei are far apart, the asymptotic density $\rho_n^{(G)}$ is still well defined, and the radius of the neutron distribution R_n in the nuclei is given by

$$\rho_n^{(N)}(R_n) = \rho_n^{(G)} \tag{5.20}$$

and the neutron number N of the nuclei is

$$N = \int_0^{R_n} \rho_n^{(N)}(\mathbf{r})\, d\mathbf{r} \tag{5.21}$$

The effects of the neutron gas are twofold: first, the proton distribution is extended beyond the neutron distribution with $\langle r_p^2 \rangle$ exceeding $\langle r_n^2 \rangle$ by ~ 20 per cent by the time $\rho \sim 10^{14}$ g cm^{-3}; second, with the reduced surface energy the nuclei are extremely susceptible to collective excitations, either vibrations or rotations of intrinsically deformed shapes. Thus even at temperatures $T \simeq 10^6$ K there may be substantial thermal energy stored in collective nuclear excitations.

We now consider the possible existence of a proton drip region. The difference between the protons and the neutrons is that a proton outside the nucleus still has a long range Coulomb interaction with the lattice and this is a positive energy increasing the chemical potential. The condition for equilibrium is

$$\mu_p^{(N)} = \mu_p^{(G)} \equiv m_p c^2 + \mu_p^{(n)} + \mu_p^{(L)} \tag{5.22}$$

The contribution $\mu_p^{(n)}$ arising from the nuclear interaction between the proton and the neutron gas is independent of position in the asymptotic region between the nuclei and depends only on the neutron gas density $\rho_n^{(G)}$. On the other hand, $\mu_p^{(L)}$ representing the electrostatic interaction of the proton with the lattice nuclei depends sensitively on the position in the unit cell. The electrostatic energy is clearly a minimum at the surface of the Wigner–Seitz cell, and, expanding the electrostatic potential about this point r_L we find that the proton experiences the electrostatic potential

$$\phi(r) = \phi_0(r_L) + \frac{3Ze}{2r_L^3}(r-r_L)^2 + \ldots \tag{5.23}$$

and hence has the zero point energy

$$\tfrac{1}{2}\hbar\omega = \tfrac{1}{2}\hbar\left(\frac{3Ze^2}{m_p r_L^3}\right)^{\frac{1}{2}} \tag{5.24}$$

in addition to the electrostatic energy

$$e\phi_0 = \frac{3}{10}\frac{Ze^2}{r_L}(1 - r_N^2/r_L^2). \tag{5.25}$$

The result is that the proton drip region is postponed relative to the onset of neutron drip.

As the density increases so does the proportion of the neutrons in the gas relative to those in the nuclei; the nuclei grow softer and softer with the proton distribution in the nuclei extending more and more beyond the neutron distribution. Just before proton drip sets in, the proton distributions of neighbouring nuclei begin to overlap, and the neutron distributions merge smoothly into the neutron gas. The nuclei have completely dissolved into a β-stable nucleon fluid. The transition occurs smoothly at $\sim 2.0 \times 10^{14}$ g cm^{-3}.

In Figs. 5.1 to 5.3 we summarize the conclusions of §5.2 and §5.3. We shall discuss the actual form of the equation of state illustrated in Fig. 5.1 more fully in Chapter 6. We point

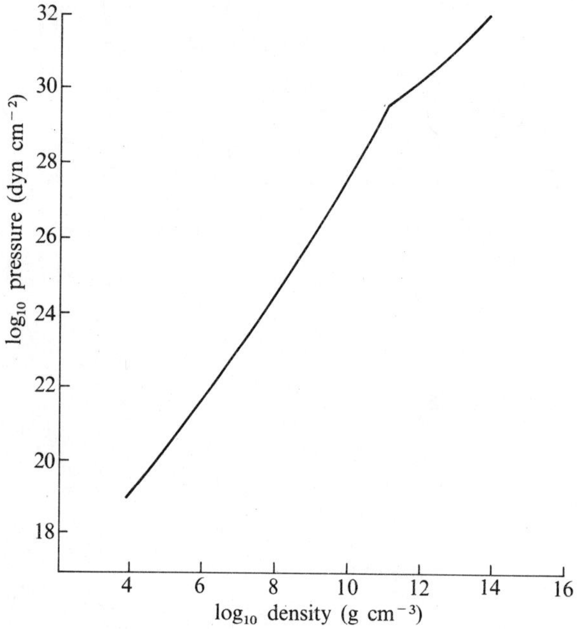

Fig. 5.1. the equation of state for neutron star matter in the crust region, exhibiting a first-order phase transition at 2×10^{11} g cm^{-3} corresponding to the onset of the neutron drip region. Taken from the papers by Baym, Pethick, and Sutherland and Baym, Bethe, and Pethick.

out in passing that here we are dealing with very exotic nuclei, and a major uncertainty is probably the calculation of the energy of the nuclei. For example, just prior to the neutron drip onset we are predicting the existence of nuclear species such as ^{106}Ni.

Before concluding our discussion of the neutron star crust, we comment briefly on a possible mechanism for the production

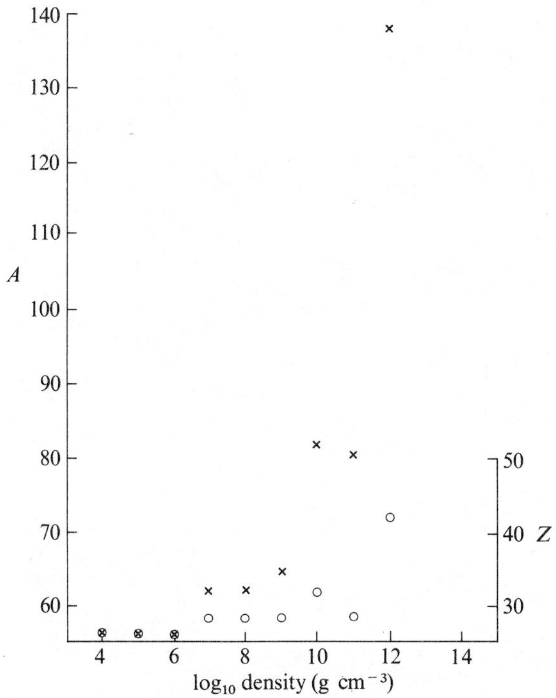

Fig. 5.2. Atomic mass numbers (X) and charge numbers (O) of most probable nuclear species as a function of density in the neutron star crust region. Note the dip in A and Z associated with the neutron drip phase transition.

of extremely heavy nuclei utilizing the exotic nuclear species found in the neutron star crust.

In Chapter 1 we reviewed the role of fusion in main sequence stars leading to the formation of nuclei as heavy as iron. Beyond iron the fusion process becomes endothermic. Beyond this point the dominant process leading to the production of heavier elements is thought to be the rapid absorption of neutrons followed by beta decay, known as the r-process. If nuclei are bathed in a sufficiently high neutron flux such that the neutron capture rate exceeds the beta decay rate, then heavier chemical species will be generated. Detailed calculations suggest that such a process can account for the relative abundances of elements as heavy as lead, but falls short in predicting the abundances of very heavy elements like uranium.

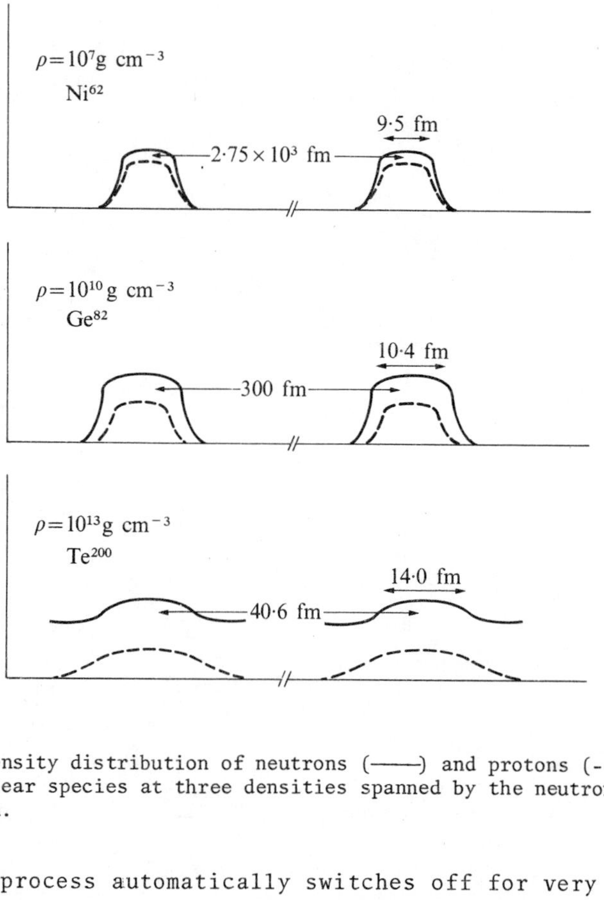

Fig. 5.3. Density distribution of neutrons (——) and protons (----) for typical nuclear species at three densities spanned by the neutron star crust region.

Such a process automatically switches off for very heavy nuclei where the dominant decay mode is by fission. Since the neutron-induced fission rates increase with the neutron flux there clearly comes a point where no new chemical species will be produced by the r-process, and calculations show that elements beyond $Z = 100$ are unlikely to be produced by such a process. Of course the elements between uranium ($Z = 92$) and fermium ($Z = 100$) are so unstable that even were they produced they would not be naturally occuring on earth.

In recent years searches for superheavy elements have been undertaken and theoretical calculations have suggested that, once created, elements like $^{354}126$ may have extremely long lifetimes ($\sim 10^9$ years). If such elements do exist, then the question must be raised, how could they be created? None of the conventional nucleosynthesis processes could account for them.

The most obvious alternatives to the r-process for the formation of heavy nuclei are heavy ion fusion reactions. The problem is that in order to obtain a sizeable fusion cross-section the neutron-to-proton ratio in the target–projectile system should be similar to that in the final fused nucleus. Since the neutron excess increases with increasing charge numbers, this is usually impossible to achieve with beta-stable nuclei. To illustrate this point consider the formation of ^{238}U. The most relatively neutron-rich stable isotope available on earth is ^{48}Ca; a partner with which this could fuse to form ^{238}U would be ^{190}Hf, while the heaviest isotope of this element available on earth is ^{183}Hf. Or consider the fusion partner for ^{56}Fe to form ^{238}U; it would be ^{182}Dy, while on earth the heaviest known isotope of this element is ^{167}Dy. The situation becomes much more difficult if we consider the formation of superheavy elements like ^{354}126. In Table 5.1 we present pairs of nuclei which could act as fusion partners for such a nucleus.

TABLE 5.1

Fusion partners in the formation of ^{354}*126*

Nucleus	Fusion Partner	Heaviest isotope of fusion partners so far observed
^{48}Ca	^{306}106	^{264}106
^{56}Fe	^{298}Fm	^{258}Fm
^{208}Pb	^{146}Ru	^{109}Ru
^{238}U	^{116}Se	^{87}Se

We note that the exotically neutron-rich fusion partners necessary for the formation of heavy elements do exist in the crust of neutron stars. However, we have argued that the temperature is unlikely to exceed $\sim 10^6$ K which is too cool for thermonuclear reactions involving these nuclides. In chapter 3 we described the importance of neutrino cooling in neutron stars and we observed (eqn. 3.19) that because of weak neutral currents the neutron star crust opacity was proportional to A^2 where A was the mass number of the nuclear species present. For $A = 50$ at a density of 10^{12} g cm^{-3} the neutrino mean free path was comparable with the crust thickness and it is correspondingly shorter for nuclei with $A \gg 50$. We thus have the possibility that large mass, exotically neutron-rich isotopes from deep within the crust, below the neutron drip region, may be driven outwards by neutrino pressure and into collisions with nuclei like ^{56}Fe nearer the surface and with which they may fuse to form heavy elements. Indeed if the neutrino pressure is great enough the crust may be blown off the star explosively thus adding to the universal abundance of very heavy elements.

5.4. The nuclear fluid

An enormous effort has gone into calculating the properties of the nuclear fluid known as 'nuclear matter'. This is a hypothetical fluid of equal numbers of neutrons and protons in which the Coulomb interaction is ignored. Nuclear matter has served as a model for the development of a large number of many-body techniques some of which are described in Appendix A. In neutron star physics we are interested in the properties of a beta-stable nucleon fluid. The condition of beta stability is given in terms of the neutron, proton, and electron chemical potentials μ_n, μ_p, and μ_e respectively as

$$\mu_n - \mu_p = \mu_e. \tag{5.26}$$

These in turn are given by the conditions that at a given density ρ there be conservation of baryon number B and charge number Q. The techniques used for nuclear matter calculations and discussed in Appendix A can be used directly in calculating the internal energy as a function of density for beta-stable matter.

Pairing correlations in the neutron and proton fluids can be explicitly taken into account as described in Appendix B. The net effect on the energy–density curve is to lower the energy by ~ 1 per cent. At a given density due to s-wave pairing this is far less than the uncertainty with which the energy–density curve can be calculated. It would thus appear that pairing correlations have little significance for the calculation of an equation of state although they are clearly of paramount importance in discussing relaxation following a crust spin-up.

At densities $\rho \geqslant 3 \times 10^{14}$ g cm^{-3} the s-wave pairing vanishes and there is some suggestion of a growth of a $^3P_2$ state pairing which would imply an anisotropic superfluid. However, accurate calculations have not been performed.

At a density $\rho = 3 \times 10^{14}$ g cm^{-3} the electron chemical potential in eqn. (5.26) is $\mu_e \simeq 100$ meV. As the density is increased so also is the value of μ_e. When μ_e exceeds 105 MeV (the rest mass of the muon), then muon production can begin and we have a dynamic stability governed by

$$\mu_n - \mu_p = \mu_e = \mu_\mu. \tag{5.27}$$

However the number of muons is extremely small, and, for densities less than 10^{15} g cm^{-3}, they play only a minor role in determining the equation of state. Indeed before the muons are of any significance a number of other phenomena have to be reckoned with, and these are discussed in the remainder of this chapter.

5.5. Neutron solidification

As we compress matter beyond nuclear saturation densities, we enter a new regime of physics not found elsewhere in nature. It is true that relativistic heavy ion reactions which have only recently become the subject of laboratory studies may well involve extremely dense states of nucleon matter. But these states are likely to differ from neutron star matter in that they may contain a lot of free energy, and it is unlikely that they will come into thermal equilibrium, and in all probability their properties will be complicated by surface effects. Nevertheless, we shall comment from time to time on the relevance of various heavy ion experiments which may yield some insight into the structure of dense matter.

What we might expect to happen as the pressure and density increase is that at various critical points matter will undergo phase transitions, but with little experience to guide us we can only speculate on the form the new phases might take. In this we can be guided by what happens in another dense Fermi fluid, namely liquid helium-3. At low pressures and temperatures there are new phases of liquid helium-3 which are anisotropic superfluids. As we have mentioned in §5.4, such effects may occur in nucleon fluid at densities $\sim 10^{15}$ g cm^{-3}, but the condensation energy, i.e. pairing gap, is likely to be small, and the critical point is extremely sensitive to the method of calculation. Little progress has been made on the question of anisotropic superfluid phases of nucleon fluids and we shall not discuss them further. More appropriate for the high pressure region in the core of a neutron star is that liquid helium-3 under pressure solidifies, and corresponding states arguments have been advanced to suggest that a similar phenomenon may occur for nucleons under pressure.

A closer examination of the differences between the helium–helium atomic potential and the nucleon–nucleon interaction leaves as uncertain whether or not the simple corresponding states argument is valid. The repulsion between the helium atoms at short distances arises from the exclusion principle between the electronic clouds of the two atoms whereas the short range nuclear repulsion arises from the exchange of heavy mesons with a wide range of quantum numbers so it is not clear that under extreme pressures and densities these interactions in the many-body media will behave in a similar fashion.

If we assume that there is some given nucleon–nucleon interaction, for example the Reid potential, which is deduced from two nucleon scattering data and fitted to the nucleon–nucleon bound state, the deuteron, then we may use the method of correlated trial wavefunctions to calculate the energy of a neutron solid and compare it with the nuclear fluid of §5.4. If we find a density above which the energy of the solid is lower than the energy of the fluid, then this is the critical density for neutron solidification.

For the fluid we consider trial wavefunctions of the form (see Appendix A)

$$\Psi_F^T = F\phi_F \tag{5.28}$$

where F contains the explicit correlations which we think most important and ϕ_F is a Slater determinant of plane waves appropriate for a homogeneous fluid. Given some hamiltonian H, the correlation functions and the single particle states in the Slater determinant are formally obtained by minimizing the expectation value of H in the set of normalized trial wavefunctions (5.28). Since this is never a free variation, various constraints have to be added as discussed in Appendix A. For the solid the calculations proceed exactly as for the fluid by considering trial functions

$$\Psi_s^T = F\phi_s \tag{5.29}$$

where ϕ_s is now a Slater determinant of gaussian wave packets on a lattice appropriate to a solid. For a given space lattice the density determines the lattice spacing. To calculate the parameters of the solid the expectation value of H in the set of trial wavefunctions Ψ_s^T is minimized with respect to the correlation functions, lattice structure, and width of the gaussian wave packets on the lattice sites. If the energy of the lattice is found to be lower than that of the fluid for gaussian packets whose width is less than the lattice spacing, we would predict that the ground state was a solid. For such calculations with the Reid interaction the predictions are that the nucleons remain a fluid up to densities $\sim 1.4 \times 10^{15}$ g cm^3.

One higher order effect not taken into account in these calculations is the role of the $\Delta(1236)$ nucleon resonance. A phenomenological interaction like the Reid interaction includes the effects of repeated scatterings in which we sum over all intermediate nucleon states (see Fig. 5.4) including excited states in which the nucleon becomes a Δ. In carrying out many-body calculations in which we develop an effective interaction (say in Brueckner theory) we include the many-body effects, principally the Pauli principle and production by the background of these intermediate states. If we specifically allow one or more of the intermediate states to be a Δ, we find that this has the effect of raising the critical density at which nucleon matter will solidify.

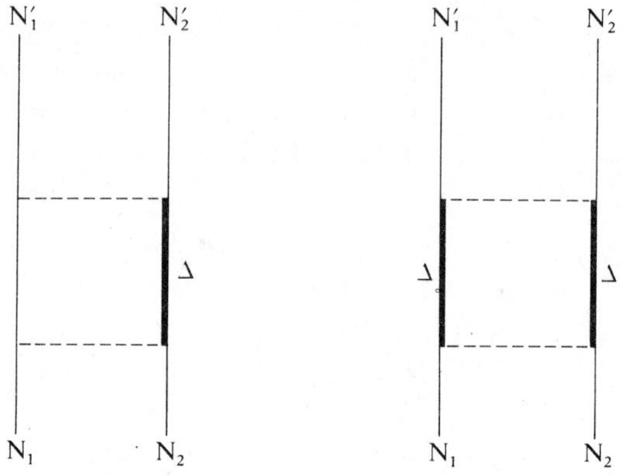

Fig. 5.4. Typical diagrams for the scattering of a pair of nucleons in which one or both nucleons is excited into the 3-3 resonance or Δ(1234 MeV) in an intermediate state.

5.6. Pion condensation

In beta-stable matter we are dealing with a nucleon fluid containing electrons and which is in thermal equilibrium under the reaction

$$n \leftrightarrow p + e + \bar{\nu}_e, \tag{5.30}$$

the condition for equilibrium being the equality (5.26). As the density increases (see §5.4) we have the possibility of muon production with the condition for equilibrium becoming (eqn. (5.27))

$$\mu_n - \mu_p = \mu_e = \mu_\mu \tag{5.31}$$

and we have thermal equilibrium of the reactions

$$n \leftrightarrow p + \mu + \bar{\nu}_\mu \tag{5.32}$$

The muons, note, are dynamically stable. The decay

$$\mu \rightarrow e + \bar{\nu}_e + \nu_\mu \tag{5.33}$$

is energetically forbidden by the electron Pauli principle.

In order of ascending mass the next negatively-charged particle is the π^- at 138 MeV rest mass and when

$$\mu_n - \mu_p = \mu_\pi \tag{5.34}$$

we will seek thermal equilibrium under the reactions

$$n \to p + \pi^- . \tag{5.35}$$

However, the pions differ from the electrons and the muons in a number of vital respects. First, the pions are hadrons, i.e. strongly interacting particles, and in a dense baryonic background their properties may be strongly renormalized. In particular the effective mass of the pions may be considerably reduced below the free mass value and this means that the neutron–proton chemical potential difference need not be as great as 138 MeV before the equilibrium condition (5.34) is met. Second, the pions are bosons unlike the electrons and muons which are fermions, and at low temperatures we would expect a bose condensate of pions, i.e. the pions will condense to macroscopically populate a single state. Finally, the pion–nucleon interaction is strongly spin and momentum dependent and this will yield a bose condensate unlike any of the other bose condensates so far studied, e.g. superfluid helium-4.

If we consider only the pion–nucleon s-wave interaction the self-energy is

$$\begin{aligned}\Sigma_\pi &= -\frac{2\pi h^2}{m_\pi}\left[a_{3/2}\rho_n + \frac{1}{3}(2a_{1/2}+a_{3/2})\rho_p\right] \\ &\simeq 219(\rho_n-\rho_p)\ \text{MeV}\end{aligned} \tag{5.36}$$

where $a_{3/2}$ and $a_{1/2}$ are the isoquartet and isodoublet π–N s-wave scattering amplitudes. Thus the threshold for pion production is

$$\mu_\pi = \mu_n - \mu_p = m_\pi + \Sigma_\pi \simeq m_\pi + 219(\rho_n-\rho_p). \tag{5.37}$$

As we increase the baryon density to increase $\mu_n - \mu_p$, we also increase $\rho_n - \rho_p$ and in fact the pion self-energy always rises faster than the difference in the chemical potentials so that if there was only a π–N s-wave interaction, pion production would never become energetically favourable. However we know that the π–N interaction is dominated by the attractive p-wave interaction. The nonrelativistic reduction of this interaction is

$$V_{\pi N}^{(p)} \sim \boldsymbol{\sigma}\cdot\mathbf{k} \tag{5.38}$$

where σ is the nucleon spin and $\mathbf{k}$ is the pion momentum. If there is a bose condensate of pions, then they will all have the same momentum $\mathbf{k}$, indeed the same four momentum $(\mu_\pi, \mathbf{k})$. Thus the energy of the pions will be of the form, kinetic energy plus potential energy

$$E_\pi(\mathbf{k}) \sim ak^2 - \mathbf{b}\cdot\mathbf{k} \tag{5.39}$$

and for the ground state

$$\frac{\partial E_\pi}{\partial \mathbf{k}} = 0 \tag{5.40}$$

which occurs for $\mathbf{k} \neq 0$. Thus this particular bose condensate will be charged and have a non-zero momentum, and for $\mu_\pi \neq 0$ it will form a supercurrent. We note that in the usual discussion of bose condensates in which it is assumed that $\mathbf{k} = 0$, it is an elementary problem in statistical physics to demonstrate that $0 \leqslant \mu \leqslant m$. This is not true for the pion condensate which has non-zero momentum where we now have $0 \leqslant \mu_\pi \leqslant (m_\pi^{\ 2} + k^2)^{\frac{1}{2}}$.

The problem now is to write down a hamiltonian for a system of nucleons, pions, and leptons and, at a fixed mass density, to minimize this hamiltonian with respect to the pion amplitude and condensate momenta in order to find the ground state. This minimization has to be carried out subject to conservation of baryon number density $\bar{N}N$, and charge neutrality

$$\{N^{+}\tfrac{1}{2}(1+\tau_3)N + P_{\pi_1}\pi_2 - P_{\pi_1}\pi_1 - \psi_e^{+}\psi_e - \psi_\mu^{+}\psi_\mu\} = 0 \tag{5.41}$$

where π_1, π_2 are the first two components of the pion isovector and $\mathbf{P}_\pi$ is the conjugate momentum to the pion field.

There are many variations on the form of hamiltonian considered and we shall simply quote the σ-model as an example

$$\begin{aligned} H_0 &= i\bar{N}\gamma\cdot\nabla N + g_{\pi NN}\bar{N}(\sigma + i\boldsymbol{\tau}\cdot\boldsymbol{\pi}\gamma_5)N \\ &+ \tfrac{1}{2}b(\sigma^2+\boldsymbol{\pi}\cdot\boldsymbol{\pi})^2 - a(\sigma^2+\boldsymbol{\pi}\cdot\boldsymbol{\pi}) - f_\pi \pi^2\sigma \\ &- i\bar{e}\gamma\cdot\nabla e + m_e\bar{e}e - i\bar{\mu}\gamma\cdot\nabla\mu + m_\mu\bar{\mu}\mu. \end{aligned} \tag{5.42}$$

Here the constants a and b are given by

$$a = \tfrac{1}{2}(m_\sigma^2 - 3m_\pi^2)$$

$$b = \frac{m_\sigma^2 - m_\pi^2}{4f_\pi^2} \tag{5.43}$$

where f_π is the pion decay constant with a value $\simeq$ 95 MeV as deduced from $\pi^\pm \to \mu^\pm$, ν reactions. The mass of the σ meson is not known but usually it is assumed that $m_\sigma \gtrsim 5$ GeV. The electromagnetic interaction is easily accounted for by adding the kinetic energy term $-\tfrac{1}{4}(F^{\mu\nu})$ to (5.42) and replacing charged particle momenta p^μ by

$$p^\mu \to p^\mu + eA^\mu. \tag{5.44}$$

It is next customary to assert that, since we are looking for a ground state containing a pion condensate, we may replace the meson field operators in the hamiltonian with their ground state expectation values and seek a minimum with respect to these c-numbers. This approximation however precludes us iterating the hamiltonian to generate pion exchange contributions to the nucleon–nucleon interaction, and in its approximate form H_0 contains no nucleon–nucleon interaction term. It is therefore necessary to add such a term $E_N(\rho)$ to H_0. The form $E_N(\rho)$ is usually that derived for the nuclear fluid or nucleon solid regions discussed in §5.4 and §5.5. There remains however another omission; the nucleon–nucleon interactions produce short range correlations between the nucleons (see Appendix A). The problem of propagation through a correlated medium was first discussed by Lorenz in the 1870s, and in the present context it is easy to see physically what will happen. When a pion interacts with a nucleon the repulsive core of this nucleon will repel nucleon neighbours so that the number of nucleons the pion interacts with simultaneously will be reduced and hence the magnitude of the pion self energy will be reduced, i.e. the effective mass will increase and the onset of pion condensation will be postponed to higher densities. Finally, if we are in a region of pion condensation why should the π–N resonance $\Delta(1234)$ not become dynamically stable? If this happens, then we would expect the number density of Δs to increase much faster than the nucleon number density as the density increases. Since Δ is an isospin-3/2, spin-3/2 object, each momentum state is sixteen-fold degenerate whereas each nucleon state is at most four-fold degenerate. Of course, once pions can be spontaneously produced we lose isospin and parity (the quantum numbers of the pion) as conserved quantities. The loss of parity is already evident in the existence of a non-zero condensate momentum $\mathbf{k}$ so that the system is not reflection invariant. The inclusion of Δs in the σ-model of (5.42) adds an additional term

$$H_\Delta = -i\bar{\Delta}\gamma\cdot\nabla N + g_{\pi NN}\bar{\Delta}(\sigma+i\boldsymbol{\tau}\cdot\boldsymbol{\pi}\gamma_5)\Delta + \Delta M\bar{\Delta}\Delta \qquad (5.45)$$

to the hamiltonian where $\Delta M \sim 2m_\pi$ is the mass difference between free nucleons and free Δs. The additional interaction between the pions and the Δs increases the magnitude of the pion self-energy, i.e. decreases the pion effective mass. Most calculations suggest that the effect of the short range nucleon correlations and the Δ–π interactions cancel each other out as far as the critical density for the onset of pions is concerned.

Because of the momentum dependence of the pion–nucleon interaction, e.g. see eqn (5.38), se see that, as the pion amplitude increases, the energy can decrease without limit with increasing condensate momentum. However the free pion–nucleon scattering cross-section falls off with increasing energy faster than a point $\boldsymbol{\sigma}\cdot\mathbf{k}$ interaction would suggest (see Fig. 5.5) and, to

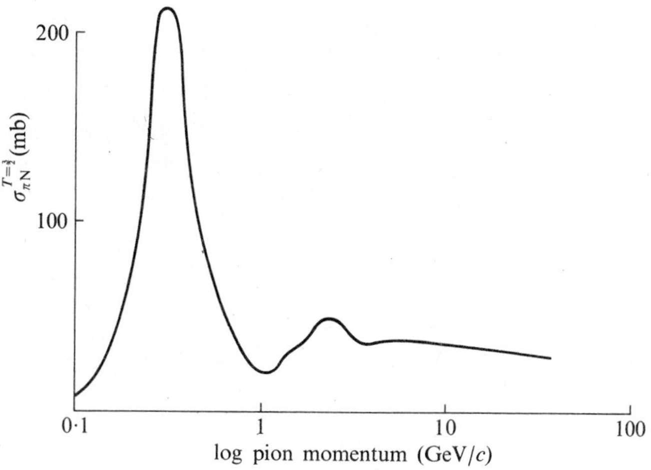

Fig. 5.5. The π–n scattering cross-section. Note that it tends to a constant at large momenta whereas the usual p-wave interaction would lead to an increasing cross-section.

include this effect in our model, it is usual to give the pions a form factor corresponding to a momentum cut-off $\gtrsim$ 1 GeV/c. This masks our uncertainty in the higher partial wave π–N interactions which become of increasing importance as the density increases.

When the various features of the calculation are put together we come up with the prediction that pion condensation should begin at $\sim 3 \times 10^{14}$ g cm^{-3}.

There is, however, one further twist to the story. Since we have lost isospin as a conserved quantum number it seems pointless to be talking in terms of neutrons, protons, and Δs. If instead we diagonalize the baryonic part of the hamiltonian $H_0 + H_\Delta$ we find that we have 10 doubly-degenerate (in z-component of spin time reversed pair states, i.e. σ and $-\sigma$) baryon Fermi seas. These decompose into four spin component-3/2 seas formed from Δ isospin mixtures and six spin component-1/2 seas formed from nucleon–Δ isospin mixtures. Only one of these seas has negative free energy in the ground state, i.e. we have a single degenerate Fermi sea of quasiparticles occupied in the ground state. These quasiparticles are linear combinations of nucleon and Δ states being mostly neutrons with admixtures of protons and Δs. The increased spin degeneracy of Δ-states makes the Δ–N spin-dependent interactions stronger than the N–N spin-dependent interactions and in particular the tensor force which is propagated by one-pion exchange. We have already seen (eqn. (5.38)) that there is a strong coupling between the baryonic spin and the condensate momentum. It then follows that a spin-ordered lattice of baryonic quasiparticles has a lower energy than the baryonic fluid (see Fig. 5.6). Within this lattice the

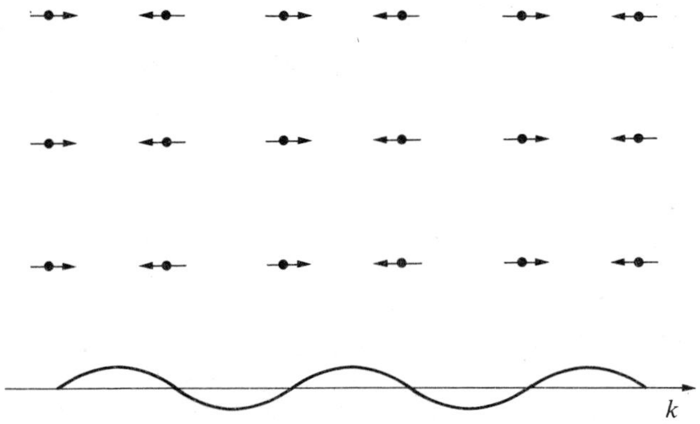

Fig. 5.6. The baryonic layered spin lattice associated with a pion condensate of momentum k.

pion condensate plays the role of a spin–isospin wave, analogous to the familiar spin wave in more conventional solids. In Fig. 5.7 we plot the calculated baryonic abundances as a function of density. We shall comment on the relevance of these to equations of state in §6.1.

The mechanical and electrical properties of such a system have not been fully explored. Clearly the presence of pions will have a marked effect on the neutron star cooling rate (see §3.1).

While we have discussed pion condensation within the framework of the σ-model, it should be stressed that similar results can be obtained with various assumptions about the form of the starting hamiltonian provided only that it adequately represents the well known π–N p-wave interaction.

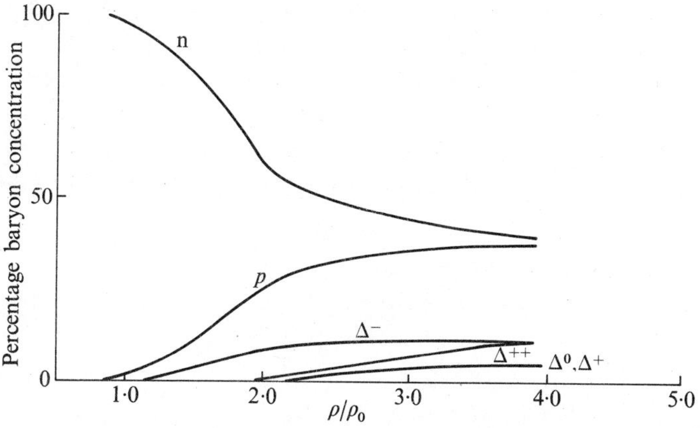

Fig. 5.7. Percentage baryon concentration in the presence of pion condensation as a function of density ($\rho_0 = 2.8 \times 10^{14}$ g cm^{-3}).

Thought has also been given to the possibility of a π^0 condensate. Being uncharged, there is no chemical potential for π^0s, and they would condense on the four-momentum state $(0,\mathbf{k})$, i.e. a static standing wave. The presence of such a standing wave would imply a baryonic lattice in the same way that the measurement of a periodic Coulomb field would imply a periodic charge distribution as a source of the field. However, such a static field is fully accounted for in the N–N interaction and there is no condensate energy, i.e. a phenomenological interaction should predict the associated solidification, if it is going to occur without specific reference to the pion fields.

5.7. Hadronic cores

In the previous section we discussed the possible role played by pions and $\Delta(1234)$s in the structure of dense neutron star matter, but clearly these are just the forerunners of a whole series of hadrons which may enter as the density increases. Indeed if we ignore interactions we would expect the $\Lambda^0(1116)$ and $\Sigma(1195)$s to make their appearance earlier than the $\Delta(1234)$, and, of course, there will also be an increasing number of meson species presumably forming a hierarchy of condensates. As the density increases, the higher the spin degeneracy of a species the more rapidly will the number density of that species rise relative to lower spin species because of the degeneracy factor in the density of states expression for a Fermi gas.

The problem in dealing with hadronic matter is that we are now beset with all the problems of high energy physics. In the earlier sections of this chapter we have had as input the data from low energy nuclear physics, and, even if this is not fully understood at the level of a unified underlying field theory, there is a great deal of phenomenology that can be used as a guide. Even when discussing pion condensation we require only a modest extrapolation from well tabulated pion—nucleon phenomenology. As the mass of hadrons considered increases, the phenomenology becomes less well substantiated. The problem does not end here for even if we believed that the interactions between the free hadrons were understood the many-body effects certainly are not.

Arguments regarding the nature of the interactions which should be involved include:

(a) At densities greater than $\sim 10^{15}$ g cm^{-3} the hadrons are so close together that all the baryons see a common repulsive interaction, presumably generated by exchange of members of the ω-meson family.

(b) While the ω-mesons provide a repulsion there will be an attractive interaction provided by the f-meson family and that as a result baryon—baryon interactions may be ignored.

(c) Repulsive interactions will serve to eliminate possible resonances; however, this elimination is limited by causality. On the other hand attractive interactions can generate a multiplicity of resonances. Since there is no limit to the rate of increase with energy of the number of states which are so generated, then, provided a channel exists which will provide an attractive interaction, there will be an exponential increase in the density of states which will swamp any effect of repulsion in other channels.

We will limit our comments to a brief discussion of the Hagedorn statistical bootstrap approach to the density of hadronic states. The basic feature of the approach is the bootstrap hypothesis, i.e. the hadrons are themselves composed of hadrons, the interactions between which generate a self-consistent observable free hadronic spectrum. The whole system is assumed to be in statistical equilibrium. This implies that if we assume a density of states $\rho_{\rm in}(\varepsilon)$ for the particles of total energy ε and use this to calculate the density of states $\rho_{\rm out}(\varepsilon)$, then

$$\rho_{\rm out}(\varepsilon) = \rho_{\rm in}(\varepsilon) \tag{5.46}$$

The number of states of one particle of spin degeneracy g in a box of volume V with momentum between $\mathbf{p}$ and $\mathbf{p} + \mathbf{dp}$ is $gV\mathbf{dp}/h^3$ and generalized to n independent particles with total energy ε is

$$\rho_n(\varepsilon) = \left(\frac{V}{h^3}\right)^{n-1} \prod_{i=1}^{n} g_i \int \mathbf{dp}_i \delta\left(\sum_{i=1}^{n} \varepsilon_i - \varepsilon\right) \delta^3\left(\sum_{i=1}^{n} \mathbf{p}_i\right) \tag{5.47}$$

where the momentum delta function and the missing factor (V/h^3) correspond to taking the centre of mass at rest. Further generalized to an assumed density of states $\rho_{\rm in}(\varepsilon)$ for the bootstrap hypothesis we have

$$\rho_{\rm out}(\varepsilon) = \sum_{i=2}^{n} \frac{1}{n!}\left(\frac{V}{h^3}\right)^{n-1} \prod_{i=1}^{n} \int \mathbf{dp}_i \int d\varepsilon_i \rho_{\rm in}(\varepsilon_i)$$

$$\delta\left(\sum_{i=1}^{n} \varepsilon_i - \varepsilon\right) \delta^3\left(\sum_{i=1}^{n} \mathbf{p}_i\right) \tag{5.48}$$

where the $\frac{1}{n!}$ term prevents double counting. The solution to eqns. (5.46) and (5.48) is of the form

$$\rho = c\varepsilon^a \exp(\varepsilon/\varepsilon_0). \tag{5.49}$$

If we restrict ourselves to hadrons of particular baryon number B, charge Q, strangeness S, etc., then the parameters c, a, and ε_0 are unique and are given by

$$-7/2 \lesssim a \lesssim 5/2$$

$$m_\pi \lesssim \varepsilon_0 \lesssim 174 \text{ MeV}$$

$$c \sim (0.9 \text{ GeV})^{3/2}, \quad a = -5/2$$

$$c \sim (0.2 \text{ GeV})^2, \quad a = -3. \qquad (5.50)$$

In Fig. 5.7 we plot the number densities of all known baryons with mass less than 1.7 GeV as a function of mass density obtained in a statistical bootstrap calculation by Leung and Wang.

There is a major omission in the approach described above and that is that eqn. (5.49) takes no account of the Pauli exclusion principle for the baryons. While this may not be a serious deficiency in the high energy distribution tail for hot baryon systems, it is certain to cause problems for extremely low temperature degenerate neutron star cores. We can see one example of a disagreement between the present calculation and our discussion of pion condensation in Fig. 5.8, from which we see that in the spirit of the statistical bootstrap there is no mass density for which a π^- condensate will appear. Similarly, ignoring the interactions leads to the prediction that

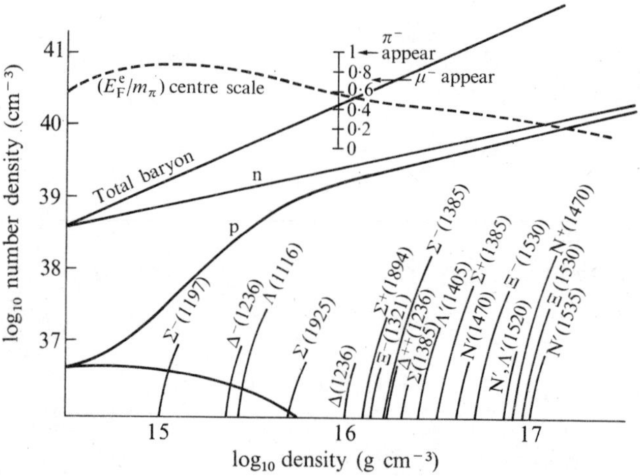

Fig. 5.8. Particle abundances in a neutron star core as predicted by a statistical bootstrap calculation. Note the centre scale and the prediction that in this model it is never energetically favourable to produce pions.

Σs will appear before Δs as discussed in the introduction to this section.

Until the question of the exclusion principle and the nature of the hadronic interactions have been resolved, the uncertainties in the statistical bootstrap approach must remain.

More recently doubt has been expressed as to whether or not the baryons can retain any identity at high densities. An alternative to the bootstrap approach to hadronic spectra has been the development of quark models which have had great success in explaining many features of hadronic spectra and of strong interaction decays. Further evidence for the suggestion that hadrons are composed of point-like fractionally charged spin-½ quarks comes from the deep inelastic scattering of leptons by baryons. Models have been proposed in which the interaction between the quarks at extremely short distances is very weak and is large at long distances. In such a model the hadrons are composed of essentially free quarks which are nevertheless constrained to remain within a volume of the size of the hadron, and no free quarks are observed. Now, since the size of the neutron is a fraction of a fermi, the neutron can be typified by a mass density 10^{15} g cm^{-3}. Thus in neutron star cores with densities $\gtrsim 10^{15}$ g cm^{-3} the baryons will seriously overlap with one another, their individuality may become confused, and the system may evolve into a degenerate quark soup. We shall elaborate on this suggestion when discussing equations of state in Chapter 6. On these speculative points we shall end our discussion of neutron star structure.

Suggested references

BAYM, G., PETHICK, C.J., and SUTHERLAND, P. (1971). *Astrophys. J.* **170**, 299. Discusses the low density crust properties of neutron stars up to the neutron drip point.

BAYM, G., BETHE, H.A., and PETHICK, C.J. (1971). *Nuc. Phys.* **A175**, 225. Discusses the nature of matter in the density range 4.3×10^{11} g cm^{-3} (the neutron drip point) up to $\sim 5 \times 10^{14}$ g cm^{-3} (neutron solidification and pion condensation not considered).

CAZZOLA, P., LUCARONI, L., and SCARINGI, C. (1966). *Nuovo Cim.* **43**, 250. First discussion of neutron solidification.

ANDERSON, P.W. and PALMER, R.G. (1971). *Nature, Lond.* **231**, 145. The corresponding states argument for neutron solidification.

CANUTO, V. and CHITRE, S.M. (1974). *Phys. Rev.* **D 9**, 1587. In favour of neutron solidification.
TAKEMORI, M.T. and GUYER, R.A. (1975). *Phys. Rev.* **D 11**, 2696. Argues against neutron solidification — represents the consensus view at this time.
PANDHARIPANDE, V.R. and SMITH, R.A. (1975). *Nuc. Phys.* **A 237**, 507. Discuss long range order in neutron matter resulting from a π^0 condensate.
BAYM, G. and FLOWERS, E. (1974). *Nuc. Phys.* **A 222**, 29. Clearly establishes the conditions for pion condensation.
CAMPBELL, D., DASHEN, R., and MANASSAH, J. (1975). *Phys. Rev.* **D 12**, 1010. Gives an elegant discussion of pion condensation as a spontaneous breaking of chiral symmetry.
IRVINE, J.M. (1975). *Rep. Prog. Phys.* **38**, 1835. Attempts to review the status of pion condensation.
COLLINS, J.C. and PERRY, M.J. (1975). *Phys. Rev. Lett.* **34**, 1353. Suggests that dense neutron stars may have quark cores.

6

NEUTRON STAR EQUATIONS OF STATE

6.1. Equations of state

THE simplest equation of state with any relevance to neutron star physics is that for a non-interacting degenerate Fermi gas of neutrons. In the nonrelativistic, low density regime [neutron Fermi energy $\ll$ neutron rest mass] with N neutrons occupying a volume V we have

$$E(N,V) = \frac{3\hbar^2 k_F^2}{10m_n} N \tag{6.1}$$

where the Fermi momentum is

$$k_F = (3\pi^2 n_n)^{1/3} \tag{6.2}$$

and the neutron number density n_n and mass density ρ are related

$$n_n = N/V = \rho/m_n. \tag{6.3}$$

The resulting pressure is then given by

$$P = -\left.\frac{\partial E}{\partial V}\right|_N = -\frac{\partial}{\partial V}\left[\frac{3\hbar^2}{10m_n}(3\pi^2)^{2/3} \frac{N^{5/3}}{V^{2/3}}\right]$$

$$= \frac{\hbar^2}{5m_n^{8/3}}(3\pi^2)^{2/3} \rho^{5/3} \tag{6.4}$$

(see eqn. (1.19)).

At very high densities we may use the extreme relativistic approximation

$$E(NV) = \hbar c k_F N \tag{6.5}$$

whence

$$P = \tfrac{1}{3}\hbar c\ m_{\mathrm{n}}^{-4/3}(3\pi^2)^{1/3}\rho^{4/3} \tag{6.6}$$

(see eqn. (1.26)).

More precisely, for the non-interacting Fermi gas we should calculate the pressure throughout the whole density range from the dispersion relation

$$E(N,V) = \frac{V}{\pi^2\hbar^3}\int_0^{\hbar k_{\mathrm{F}}} [\{p^2c^2 + m_{\mathrm{n}}^2c^4\}^{1/2} - m_{\mathrm{n}}c^2]p^2\mathrm{d}p. \tag{6.7}$$

However, since the non-interacting neutron gas is an over-simplification, we shall now move on to more realistic calculations.

For realistic calculations of the equation of state we require the numerical results of the energy density calculations for the various regions discussed in Chapter 5. The exact predictions then depend on the details of the individual calculations. However, at least up to nuclear saturation densities, there is a surprising amount of agreement on the predicted equation of state.

First we note that if the pressure rises as a function of density faster than the internal energy density ε, ultimately we shall violate the limit of causality when the velocity of sound V_{s}

$$V_{\mathrm{s}} = (\partial P/\partial\varepsilon)^{1/2} \tag{6.8}$$

exceeds the velocity of light in vacuo. Note that there is no reason why the speed of sound should not exceed the velocity of light in the neutron star matter.[1] The non-interacting relativ-

[1]The velocity of sound is determined by the equation of state (eqn. (6.8)). At neutron star densities this is dominated by the baryonic Fermi pressures and the effects of the strong nuclear interactions. In this situation the equation of state can become very stiff, and the velocity of

istic neutron gas equation of state (6.6) sets a limit on the speed of sound as

$$V_s \lesssim \frac{1}{\sqrt{3}}c. \tag{6.9}$$

Next we note that all nucleon–nucleon interactions which fit two nucleon phenomenology predict that the interacting neutron gas will have a softer equation of state than the non-interacting neutron gas for densities $\rho \lesssim 5 \times 10^{14}$ g cm^{-3}. Any phase transition that is predicted to favour a phase other than the neutron gas will lead to a further softening of the equation of state. Thus we may look upon the non-interacting neutron gas equation of state relations (6.4) and (6.6) as providing an upper limit on the pressure at a given density at least up to $\sim 5 \times 10^{14}$ g cm^{-3}. It is probably an upper limit even at higher densities because, although there must be uncertainty in treating the interactions in this density regime, we have already indicated that a rich collection of new phases is likely.

In Fig. 6.1 we reproduce some neutron gas equations of state, and in Fig. 6.2 we present an overall view of the equation of state of cold, electrically neutral matter over an extremely wide range of densities.

Returning to the specific problem of the neutron star equation of state and starting in the low density crust and working our way inwards, we find that as the density increases the first feature of note is a first-order phase transition corresponding to the onset of the neutron drip region (§5.3). Different calculations place this transition density at between 2 and 4.5×10^{11} g cm^{-3} corresponding to a pressure of between 2 and 8×10^{29} dyn cm^{-2}. The next point of interest is the density at which the nuclei dissolve into a uniform nucleon fluid (§5.4). This occurs at densities between 1 and 2.5×10^{14} g cm^{-3} by which point the pressure has risen to between 1 and 2×10^{33} dyn cm^{-2}. We next have a phase transition involving

---

sound can approach the velocity of light in vacuo. On the other hand the velocity of light in the material is determined by the refractive index for the stellar material which in turn depends on the plasma frequency for the free charges and hence on the free charge density. These can be high for neutron star matter, and hence the velocity of light may be substantially less than the velocity of light in vacuo.

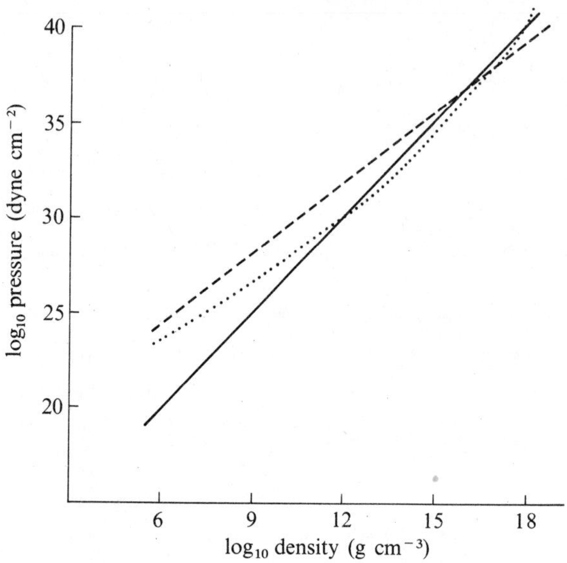

Fig. 6.1. Equations of state for pure neutron matter (——) free non relativistic neutron gas; (----) free extreme relativistic neutron gas; (····) neutron gas interacting through a static neutron–neutron potential ($^1S_0$ Reid interaction).

the appearance of a pion condensate between 3 and 6×10^{14} g cm^{-3} where the pressure has risen to $\sim 8 \times 10^{33}$ dyn cm^{-2}. Whether this is a first-order or second-order phase transition depends on the model of the pion Lagrangian.

We now comment in slightly more detail on the equation of state in the region at the onset of pion condensation. To begin with, we refer back to Fig. 5.7 where we see that, prior to the onset of pion condensation, we have almost pure neutron matter, while at higher densities we approach symmetric nuclear matter modified by the presence of Δs and the pion condensate. Some pion Lagrangians actually predict a phase transition at densities $\rho \simeq 3\rho_0$ to exact nuclear matter. In Fig. 6.3 we indicate the form of the equation of state in this region showing the transition from neutron matter to nuclear matter. The region of negative slope in the $E(\rho)$ curve is unstable. Hence in neutron stars we would expect at densities $\rho \gtrsim 3\rho_0$ to have nuclear matter with a pion condensate and Δs present; then there should be a phase boundary at which the density drops abruptly from $\sim 3\rho_0$

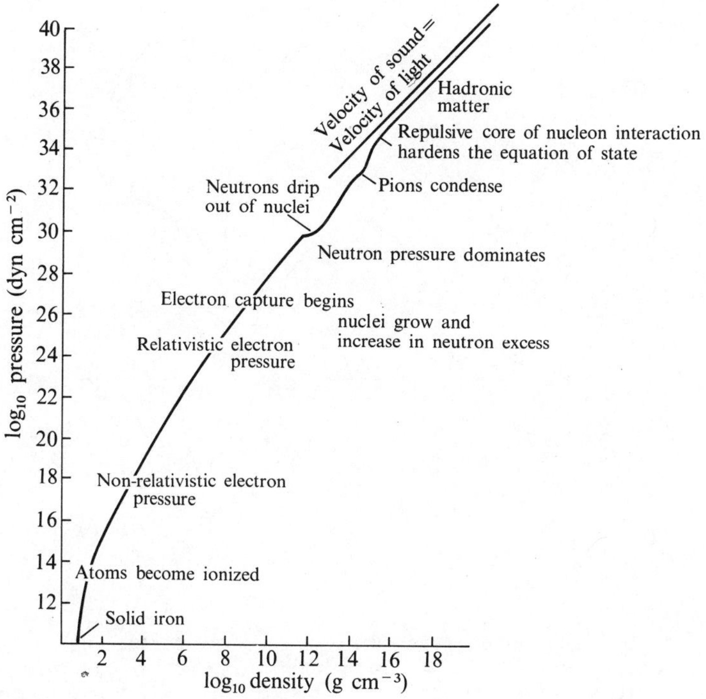

Fig. 6.2. Overall equation of state for cold matter.

to $\sim \rho_0$ and at densities $\rho \lesssim \rho_0$ we will have essentially pure neutron matter.

Beyond this density the predictions become more and more model-dependent and there is no consensus view at the present time. The only simple equation of state for the ultra-high density core is that for a non-interacting quark sea (§5.6) in which case we are back to the simple non-interacting Fermi gas equation of state.

The success of models of hadrons composed of weakly interesting quarks and yet the apparent absence of free quarks[1] has led

---

[1]At the time of writing the preliminary results of the most promising experiment designed to detect free quarks have indicated that such quarks have at last been found. Whether this result will stand the test of time remains to be seen. Even if this result does hold up, it cannot be doubted that there is a great scarcity of free quarks, and hence the discussion presented here will still remain valid.

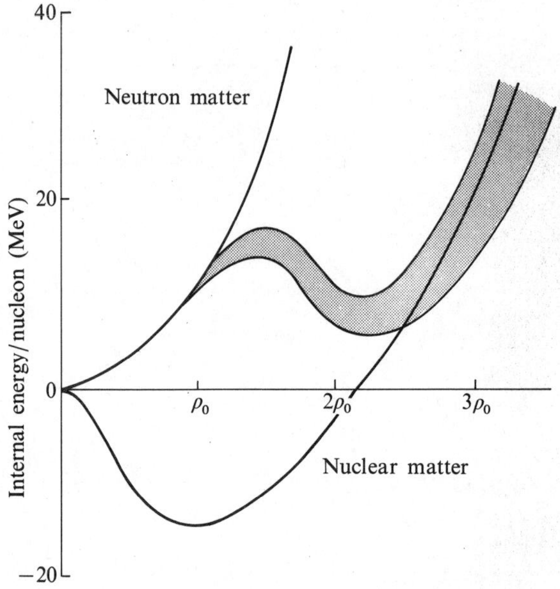

Fig. 6.3. The equation of state for neutron star matter in the region of pion condensation ($\rho_0 = 2.8 \times 10^4$ g cm^{-3}). The neutron matter and nuclear matter equations of state are shown by solid lines. The shaded region represents the probable neutron star matter equation of state.

to the development of quark confinement models in which the quarks interact with one another so as to be almost free when close together, but strongly interacting at large distances thus inhibiting the escape of quarks from the interior of the hadrons. In the most successful of such 'bag' models the spin-½ quarks have a whole new series of quantum numbers: flavoured quarks can be up, down, or strange, they can be coloured red, blue, or green and heavy quarks may exist with charm, top, and bottom. These quarks interact with one another through the exchange of members of a family of eight massless coloured 'gluons'. The lowest mass baryons and their resonances, i.e. N, Σ, Λ, Ξ, and Δ, etc. can all be constructed out of up, down, and strange quarks as can the low mass mesons, e.g. π, K, etc. Theoretical predictions suggest that the mass of the 'up' and the 'down' quark may be small (~ 0 MeV) while the strange quark is likely to be heavier with a mass ~ 300 MeV. The remaining quarks are expected to have masses ~ 1 GeV and not to appear at neutron star densities. At nuclear densities the low mass quarks are going to be highly relativistic and the simplest model for the internal energy at a quark number density n_Q would be $E = An_Q^{4/3} + B$ (see eqn. (6.6)). The parameters A and B can be fitted by minimizing the free energy $E - \mu_Q n_Q$ (μ_Q here being the quark chemical potential associated with the conservation of baryon number) and requiring that the minimum occur at density corresponding to that for a nucleon (root mean square radius ~ 0.8 fm) and at an energy E corresponding to the nucleon rest mass energy. Assuming that the quark bags have a radius ~ 0.8 fm, we expect the transition to quark matter at 3–4 times nuclear matter densities. Whether there will actually be a phase transition or the nucleons will simply dissolve (as is the case with the nuclei in the transition from the crust remnants below the neutron drip point into the nucleon fluid) is not obvious. The transition from nuclear matter to quark matter need not destroy the pion condensate. The quark matter may well be a magnetic superfluid with pairs of quarks condensing out into spin-1 states.

For a Chandrasekhar mass star with a radius ~ 10 km, the quark matter may well extend out to ~ 8 km, and as much as 80 per cent of the mass of the star could be in this phase. We summarize the conclusions of this section in Fig. 6.4 where we present our expanded view of the appropriate region from Fig. 6.2.

6.2. Neutron star masses

In the previous section we have discussed the neutron star equation of state, i.e. having calculated the internal E as a function of particle number density n. The pressure is defined to be

$$P = -\left.\frac{\partial E}{\partial V}\right|_s . \qquad (6.10)$$

The constant energy required in equation (6.10) is assured by our assumption that we are working at absolute zero temperature.

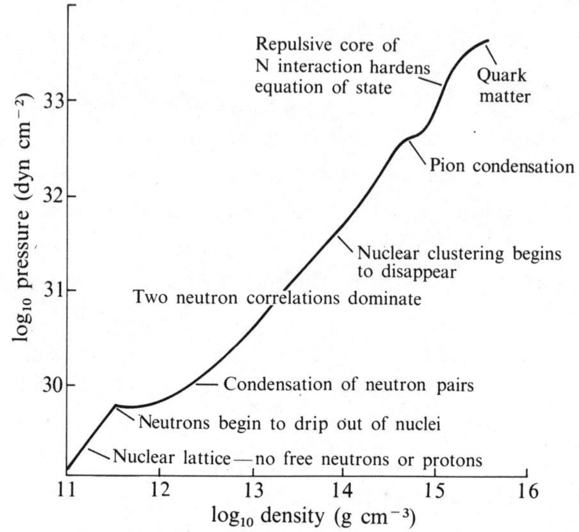

Fig. 6.4. Overall equation of state for neutron star densities.

In terms of the number density instead of volume the pressure may be written

$$P = n^2 \left.\frac{\partial \varepsilon}{\partial n}\right|_s \tag{6.11}$$

where ε is the energy per particle. In general we will have several species of particle contributing to the pressure in which case we will have a partial pressure given by eqn. (6.11) for each species and the total pressure given by the sum of the partial pressures.

Having obtained an expression for the pressure, the mass of a static spherical star is obtained by integrating the condition for stellar equilibrium (see eqn. (1.41) and Appendix C)

$$\frac{dP}{dr} = -G\frac{[M(r) + 4\pi P r^3/c^2][\rho(r) + P/c^2]}{r[r - 2GM(r)/c^2]} \tag{6.12}$$

for some central density out to the point where the pressure vanishes, having eliminated the density in favour of the pressure in eqn. (6.12) (or vice versa) by using the equation of state.

We thus obtain, for a given equation of state, a unique neutron star mass, radius and density profile for any assumed central density. Of course, all such configurations obtained by integrating the local equilibrium condition (6.12) are not necessarily stable. The maximum and minimum masses associated with a given equation of state are given by $dM[\rho(0)]/d\rho(0) = 0$. At these turning points the star has a zero frequency mode of radial oscillation, and the stability of this mode changes as we pass through the turning point. To follow this point further consider a region where the mass of the equilibrium stellar configuration is increasing as the central density increases. Take a star in equilibrium and compress it slightly; there will be a more massive star which is in equilibrium at this increased density. But the more massive star gives rise to a stronger gravitational field. Hence, our compressed star does not have a strong enough gravitational field to hold it in its new configuration, and the pressure gradients will oppose the compression. Similarly, had we rarefied the star there would have been a star of lower mass and hence lower gravitational potential, which would have been in equilibrium at the new lower density and hence the increased gravitational potential of our test star would have opposed the rarefaction. Thus, in a region where the mass of the star increases with increasing central density we have the situation where the forces engendered by a perturbation oppose that perturbation, and this is the condition for stability. Exactly similar arguments show that in a region where the mass of the equilibrium star decreases as the central density increases we have unstable equilibrium.

Plots of the equilibrium mass of cold stars as a function of central density indicate that the first maximum as the density increases occurs at $\sim 3 \times 10^8$ g cm^{-3}. At this density there is little uncertainty in the equation of state and hence the maximum mass is rather well defined and is the Chandrasekhar mass corresponding to the maximum mass of a white dwarf star. There is then a region of instability from $\sim 3 \times 10^8$ to $\sim 10^{14}$ g cm^{-3} before we have the region of stable neutron star configurations. The density of the maximum mass neutron star does depend on the equation of state, but lies in the range $\sim 10^{16}$ to 10^{17} g cm^{-3}. The maximum mass itself is extremely sensitive to the actual equation of state as we shall see.

To give us some idea of the mass limits for neutron stars we use the particularly simple, pure Fermi gas equations of state assuming that the star contains only neutrons. For very low central densities the neutron Fermi energy is always going to be very much less than the neutron rest mass energy, and the equation of state reduces to (eqn. (6.4))

$$P = \frac{\hbar^2}{5m_n^{8/3}}(3\pi^2)^{2/3}\rho^{5/3}. \tag{6.13}$$

This will correspond to the very lowest mass neutron stars for which relativistic effects will be negligible and hence we can replace the equilibrium condition (6.12) with the simpler Newtonian form (1.5) to obtain an expression for the mass of a light neutron star as a function of the central density analogous to that obtained for a black dwarf (see eqn. (1.27)) but with the electron mass replaced by the neutron mass, i.e.

$$M = M_c\left(\frac{\rho(0)}{\rho_c}\right)^{\frac{1}{2}} \tag{6.14}$$

where M_c is the Chandrasekhar mass (eqn. (1.14)) and ρ_c the corresponding central density is

$$\rho_c = 3.37 \times 10^{16} \text{ g cm}^{-3}. \tag{6.15}$$

The radius for such a star is also given by an expression analogous to that for the black dwarf star (eqn. (1.29))

$$R = 3.38\mathcal{R}_c\left(\frac{\rho_c}{\rho(0)}\right)^{1/6} \tag{6.16}$$

where $\mathcal{R}_c$ is the Schwarzschild radius for a Chandrasekhar mass star

$$\mathcal{R}_c \simeq 4.5 \text{ km}. \tag{6.17}$$

Turning now to the opposite extreme we consider the case where the density is everywhere such that the Fermi energy of the neutrons is much greater than the neutron rest mass. In this case the equation of state is given by eqn. (6.6) which, when substituted into the equilibrium condition (6.12), yields an equation for the density of the stellar matter

$$\frac{d\rho}{dr} = -4\frac{GM(r)\rho(r)}{c^2r^2}\left\{1 + \frac{4\pi r^3\rho}{3M(r)}\right\}\left\{1 - \frac{2GM(r)}{c^2r}\right\}^{-1} \tag{6.18}$$

where as before

$$M(r) = 4\pi\int_0^r \rho(r')r'^2 dr' \tag{6.19}$$

and ρ is the *relativistic* mass density

$$\rho = \varepsilon/c^2 = 3P/c^2 \tag{6.20}$$

The exact solution to eqn. (6.18) is known in the high density limit and is

$$\rho(r) = \frac{3c^2}{56\pi Gr^2} \tag{6.21}$$

Note that the relativistic mass density has a weak singularity at the origin even if the rest mass density remains finite, as it must if the Fermi velocity approaches the velocity of light. We also note that the density distribution (6.21) only goes to zero as $r \to \infty$ and there is no natural radius. Integrating equation (6.19) out to an arbitrary radius R we find

$$R = \frac{14}{3}\frac{MG}{c^2} = \frac{7}{3}\mathcal{R} \tag{6.22}$$

so that the radius R associated with a mass M is comfortably larger than the Schwarzschild radius so no catastrophic gravitational collapse is predicted, and there is no indication of a maximum mass. However, the solution (6.21) is only valid for $\hbar k_F \gg m_n c$, i.e. we should only integrate (6.19) out to a density

$$\rho \gtrsim \rho_c \tag{6.23}$$

Integrating out to ρ_c yields a maximum mass

$$M \simeq 0.5M_c. \tag{6.24}$$

Of course, the high density limit (eqn (6.21)) does not take into account the requirements of stellar stability discussed above and illustrated in Fig. 6.5 and thus does not provide us with a reliable estimate of maximum neutron star masses. However, detailed numerical integration of (6.12) with the full non-interacting neutron gas equation of state suggests a maximum mass for neutron stars close to that predicted in eqn. (6.24). This is known as the Volkov–Oppenheimer limit and plays the role that the Chandrasekhar mass does for black dwarfs. In Fig. 6.5 we plot the calculated stellar configurations for degenerate stars assuming them to be spherical, non-rotating and governed by pure Fermi gas equations of state.

There are several factors which could significantly alter the predicted maximum mass of a stable neutron star from the value given in eqn (6.24). Interactions between the neutrons will modify the pure neutron matter equation of state (see Fig. 6.1). At low densities these interactions are dominantly attractive and the equation of state will soften and the mass will decrease. However, at short distances the interaction becomes repulsive, and hence at high densities the equation of state will harden, and hence the predicted maximum mass will increase. Since the repulsive core of the nucleon–nucleon interaction has

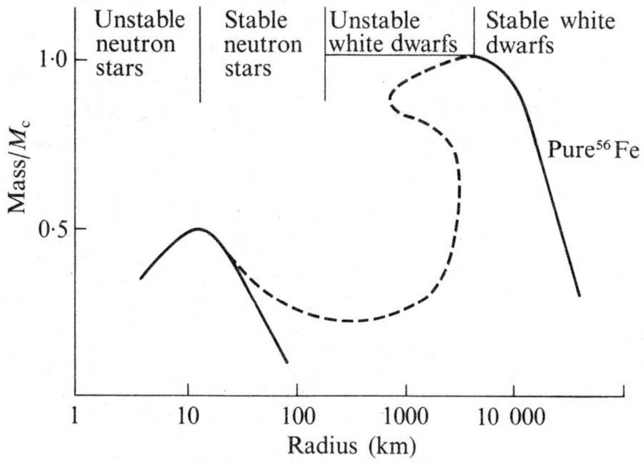

Fig. 6.5. Stellar equilibrium configurations. The solid line representing white dwarf configurations assumes a pure ^{56}Fe star supported entirely by electron degeneracy pressure; it reaches a maximum stable configuration of one Chandrasekhar mass (eqn. (1.28)). The solid curve representing neutron stars is calculated for a perfect neutron gas equation of state; it reaches a maximum stable configuration at $\sim 0.5M_C$ (eqn. (6.22)). The dotted curve is an extrapolation allowing for the chemical change from pure ^{56}Fe to pure neutrons (§1.2).

a radius ~ 0.5 fm we would not expect this to have much effect at densities below $\sim 10^{15}$ g cm^{-3}.

The introduction of interactions also introduces the possibility of phase transitions. Every phase transition implies that, at the corresponding critical density, matter can lower its free energy by adopting the new phase. Hence, every phase transition involves a softening of the equation of state and hence a reduction in the predicted maximum mass. With the possibility of pion condensation we see (§6.1) that at densities greater than $\sim 9 \times 10^{14}$ g cm^{-3} we expect to have almost pure nuclear matter (i.e. symmetric in neutrons and protons) rather than pure neutron matter. The extra degeneracy associated with isospin implies a considerable softening of the equation of state (see Fig. 6.4).

In dense matter the interaction between the nucleons is not the same as it is between two free nucleons. When we choose a phenomenological nucleon–nucleon interaction, e.g. the Reid potential, to describe free nucleon–nucleon scattering, we are including in our phenomenology repeated scattering terms including all possible intermediate states. In particular we are including amplitudes of the form shown in Fig. 5.4. When we then generate the effective interaction for use in nuclear matter (see Appendix A) using such phenomenological potentials, we are introducing an error. In the intermediate state in the amplitude shown in Fig. 5.4(a) we have a Δ and a nucleon. For free nucleon–nucleon scattering the states of these intermediate particles are unrestricted. However, when the two nucleons interact inside nuclear matter, the intermediate state nucleon is excluded from the nucleon Fermi sea by the exclusion principle. Thus, we must subtract from our interactions amplitudes like those portrayed in Fig. 5.4(a) where the nucleon is in the Fermi sea. As the density rises so does the Fermi momentum and the effect gets larger. Estimates suggest that at high densities (believed trustworthy up to 10^{15} g cm^{-3}) this could harden the equation of state by adding to the energy density a term proportional to ρ^{μ} where $\mu \sim 8/3$. Such an effect could drive up the maximum stable mass from 0.5 M_c to several times M_c. From Fig. 5.7 we see that with pion condensation the high density nuclear matter also contains Δs. Thus, we should also exclude the intermediate Δs of Fig. 5.4 from their Fermi sea which would further increase this effect. However, detailed calculations are not yet available.

Finally, rotations would add centrifugal forces to oppose the gravitational collapse and hence drive up the predicted maximum masses. Unfortunately, the solutions to Einstein's equations for a general mass distribution rotating with arbitrary angular velocities are difficult to calculate. In Appendix C we have indicated how, starting with the solution for a static spherical star, a perturbation solution in powers of the angular velocity can be obtained. The effects of rotations on stars governed by neutron matter equations of state have been studied and maximum stable masses $\sim 2M_c$ are predicted. Detailed solutions are not yet available with more realistic equations of states, but preliminary estimates indicate that maximum masses as great as $10M_c$ may not be implausible.

We see that Fig. 6.5 also predicts a minimum mass for a

stable neutron star of $0.2M_c$. The softening of the equation of state in the low density region due to the attractive nature of the neutron–neutron interaction, the extra degrees of freedom associated with beta-stability, and the phase transition to a nuclear crust will drive down this minimum mass to $\sim 0.1M_c$.

In Fig. 6.6 we plot the neutron star mass as a function of central density for the equation of state in Fig. 6.4. This

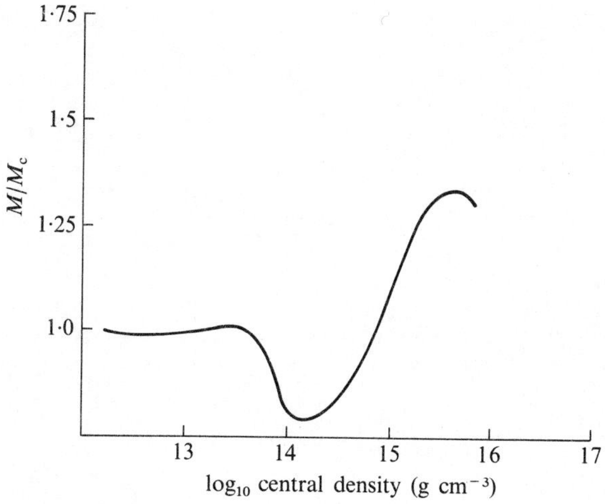

Fig. 6.6. Neutron star masses (in units of the Chandrasekhar mass as a function of the central density according to the equation of state in Fig. 6.4. Note that not all points on this curve necessarily represent stable neutron star configurations (see Fig. 6.5).

does not allow for rotations or for the hardening of the equation of state by the exclusion principle acting on the intermediate states in the amplitudes in Fig. 5.4. In Fig. 6.7 we plot the density profile for the minimum mass neutron star $M_{min} \simeq 0.1M_c$ and the maximum mass neutron star $M_{max} \simeq 1.3M_c$ predicted for the equation of state in Fig. 6.4. We see that as the mass increases the low density 'surface' region is pulled into the star. In the low mass star we clearly see the phase transition associated with the onset of neutron drip at $\sim 3 \times 10^{11}$ g cm^{-3} while in the more massive star we see the phase transition associated with pion condensation at

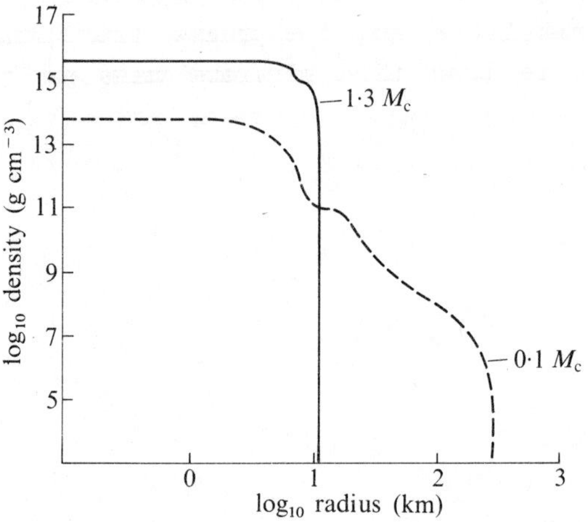

Fig. 6.7. Density profiles for neutron stars calculated from the equation of state in Fig. 6.3. As the mass of the star increases, the relatively low density crust and atmosphere are pulled into the star.

$\sim 3 \times 10^{14}$ g cm^{-3}.

6.3. Concluding remarks

There can be few physical phenomena which require for their understanding as many branches of physics as stellar structures, and neutron star structure is the example *par excellence*.

Even for the incomplete understanding that we have attempted to reach here, we have had to call on the general theory of relativity and the details of strong interaction hadron dynamics. We have used almost all of conventional nuclear physics, the physics of the solid state, and superfluid hydrodynamics. We have studied hugely distorted atoms, and been faced with problems in polymer and plasma physics. Given the nature of the problem and the wide range of tools necessary for its examination, it is perhaps not surprising that the subject is still incompletely understood.

We hope that now at least the nature of some of the problems is clear. For a more precise mass–radius relationship and upper mass limits for neutron stars we require first advances in our understanding of the equation of state for dense ($\rho \gtrsim 5 \times 10^{14}$ g cm^{-3}), strongly interacting hadronic matter. For an improved discussion of glitch phenomena we require advances in our understanding of superfluid hydrodynamics and an even more thorough analysis of neutron star crust parameters, in particular the mechanical moduli. We have not laid any emphasis on the most obvious feature of all, the source of the pulses. Here continued investigation of the magnetosphere is required so that the number of acceptable models may be reduced and a consensus of the most probable pulse source can develop.

Many stellar objects possess magnetic fields, e.g. the earth, the sun, the Ap stars, etc. There are two possible sources for such magnetic fields, either there can be some form of internal dynamo generating the fields, as is believed to be the case in the earth, or there can be 'fossil' currents set up in the object which 'freeze in' the magnetic field. Objects which have internal dynamos would have magnetic fields which would change as their structure evolves, e.g. the wanderings of the earth's magnetic pole or the violent fluctuations in the sun's magnetic field associated with sun spots. In the situation where the magnetic field is maintained by fossil currents the field will change on a time scale given by the decay constants for the currents. In the case of magnetic black dwarfs and pulsars the field is most likely due to fossil currents. The apparent stability of the magnetic fields argues for long decay times, i.e. little resistivity, for the fossil currents. A possible explanation for long decay times in black dwarfs follows from elementary solid state considerations. If the wavelength for the current carriers (here electrons) is matched to the spacing of a near perfect lattice then there is little dissipation. Such a situation could also exist in the crust region of a neutron star. An alternative and/or additional source of long-lived currents could be supplied by the supercurrents associated with the pion condensate in the core of the star.

Alternatively the magnetic superfluid quark matter core (§6.1) may undergo a ferromagnetic phase transition to provide a source for the field. Until we have developed a fully self-consistent electrodynamic model for neutron stars these questions must remain unanswered.

We have inevitably done less than justice to a number of the points raised but it is hoped that the list of references will go some way towards rectifying these shortcomings and provide the reader with an entry into the wider literature on neutron star physics.

We look forward to the development of a fully investigated self-consistent model of neutron stars which has the capacity to explain all the pulsar observations and which perhaps may give rise to suggestions for new observations to test their validity. In the development of such a model there is a rich and rewarding opportunity for a multidisciplinary application of physics to an enormous intellectual challenge.

Suggested references

Many of the references at the end of Chapter 5 discuss equations of state. In order of increasing densities the reader might like to consult

BAYM, G., PETHICK, C.J., and SUTHERLAND, P. (1971). *Astrophys J.* **170**, 299.
BAYM, G., BETHE, H.A. and PETHICK, C.J. (1971). *Nuc. Phys.* A 175, 225.
BETHE, H.A. (1971). *Rev. Nuc. Sci.* 21, 93.
SIEMENS, P. and PANDHARIPANDE, V.R. (1971). *Nuc. Phys.* **A 173**, 561.
LEUNG, Y.C. and WANG, C.G. (1971). *Astrophys. J.* **170**, 499.
COHEN, J.M. and BORNER, G. (1974). In *Physics of dense matter, IAU Symposium No. 53, 1972* (ed. C. Hansen) D. Reidel, Boston. Reviews neutron star mass-radii relationships for several equations of state.
WILSON, J.R. (1973). *Phys. Rev. Lett.* **30**, 1082. Discusses the effects of rotations on mass-radii relationships for neutron stars.
HARTLE, J.B. and THORNE, K.S. (1968). *Astrophys. J.* **153**, 807. Discusses a perturbative approach to the effects of rotations on neutron star characteristics.
OPPENHEIMER, J.R. and VOLKOFF, G. (1939). *Phys. Rev.* **55**, 374. The pioneering paper on theoretical neutron star models.
THORNE, K.S. (1965). *Science* **150**, 1671. A very clear analysis of the stability conditions in neutron stars.

APPENDIX A

CALCULATIONS OF THE BULK PROPERTIES OF DENSE FERMI SYSTEMS

THE properties of non-interacting Fermi gases, both at zero temperature and finite temperatures, are fully discussed in text books on statistical physics, e.g. Landau, L.D. and Lifshitz, E.M. (1959). *Statistical physics*, Pergamon Press, London.

At zero temperature, the wave function for the system is a single Slater determinant of plane wave states with momentum states occupied at a density of ν/h^3 per unit volume of phase space up to some Fermi momentum $\hbar k_F$ given by the number density n.

$$k_F = \left(\frac{6\pi^2}{\nu}\right)^{1/3} n^{1/3} \tag{A1}$$

where ν is the degeneracy factor associated with the internal degrees of freedom, i.e. spin, isospin etc.

If the system is weakly interacting via a two-body potential v, then we may hope that the effects of the interaction may be accommodated by a mean field description. In such a description the state of the system is still assumed to be adequately represented by a single Slater determinant, but the single particle states are chosen so as to minimize the expectation value of the hamiltonian including the interaction. Thus, if the ground state is of the form

$$\Phi(\mathbf{r}_1 \;.....\; \mathbf{r}_N) = \frac{1}{\sqrt{N!}}\det\, \phi_j(\mathbf{r}_i) \tag{A2}$$

the single particle states ϕ_i are the solutions of the Hartree–Fock equations

$$-\frac{\hbar^2}{2j}\nabla^2\phi_i(\mathbf{r}) + \sum_m \int \phi_j^*(\mathbf{r}')v(\mathbf{r},\mathbf{r}')\Big(\phi_i(\mathbf{r})\phi_j(\mathbf{r}') - \phi_i(\mathbf{r}')\phi_j(\mathbf{r})\Big)d\mathbf{r} = \varepsilon_i\phi_i(\mathbf{r}) \tag{A3}$$

which is just the Schrödinger equation for the motion of a single particle in the state ϕ_i under an average interaction with all the other particles in the system, and hence the nomenclature of a 'mean field theory'.

The total ground state expectation value of the energy is then given by

$$\langle H \rangle = \sum_i (\phi_i| - \frac{\hbar^2\nabla^2}{2m}|\phi_i) + \tfrac{1}{2}\sum_{ij}(\phi_i\phi_j|v|\phi_i\phi_j)_{\mathrm{AS}} = \tfrac{1}{2}\sum_i(\varepsilon_i + t_i) \qquad \text{(A4)}$$

where $(\phi_i\phi_j|v|\phi_i\phi_j)_{\mathrm{AS}}$ stands for the antisymmetrised matrix element

$$(\phi_i\phi_j|v|\phi_i\phi_j)_{\mathrm{AS}} = \int \phi_i^*(\mathbf{r})_j^*(\mathbf{r}')v(\mathbf{r},\mathbf{r}')\Big(\phi_i(\mathbf{r})\phi_j(\mathbf{r}') - \phi_i(\mathbf{r}')\phi_j(\mathbf{r})\Big)\mathrm{d}\mathbf{r}\mathrm{d}\mathbf{r}' \qquad \text{(A5)}$$

and t_i is the expectation value of the kinetic energy in the single-particle state ϕ_i. Hartree–Fock theory is fully reviewed in a number of text books on many-body physics, e.g. Fetter, A.L. and Walecka, J.D. (1971). *Quantum theory of many-particle systems*, McGraw Hill, New York.

If the system is strongly interacting, as for example in the case of a system of nucleons, the mean field theories break down. The reason for this is that the Hartree–Fock theory assumes that there are no two-body correlations in the system and that, despite the interactions, an individual particle model is a good approximation. The failure of this assumption is most clearly seen in the case of very strong short-range repulsions between the particles which will introduce correlations which correspond to keeping the particles apart. In classical physics, such considerations require a change in the equation of state from that of a perfect gas to that for a van der Waals gas. Other strong components in the interparticle force, for example the tensor force in the case of nucleons, may lead to other correlations.

In the case of strongly interacting fermion systems, there are historically two approaches to the problem. One is based on perturbation theory and is associated with the names of

Brueckner, Bethe, and Goldstone. The other is a variational approach and is originally associated with Jastrow.

Clearly for a strongly interacting system, we must expect to have to work to very high orders in perturbation theory. Indeed, for hard core interactions we need to work through infinite order. The repeated scattering amplitude in Fig. A1 can be computed from the Brueckner integral equation

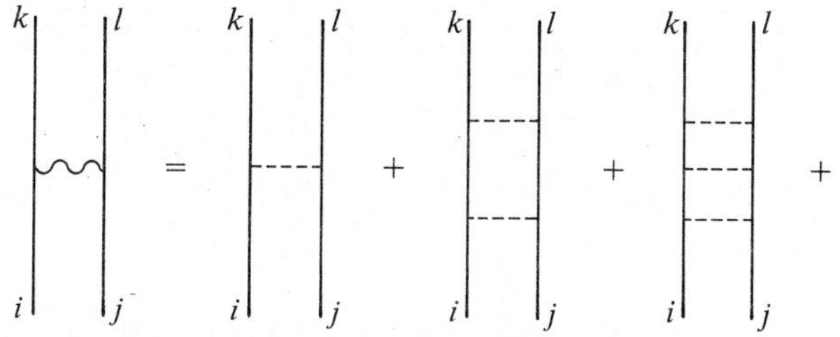

Fig. A1. The Brueckner repeated scattering amplitude describing the scattering of a pair of particles from the initial state $|ij\rangle$ to the final state $|kl\rangle$ is pictorially represented as the vertex of the wavy line and is the sum of the ladder diagrams when each dotted vertex represents the action of the bare potential v.

$$\begin{aligned} g &= v - vQ/Ev + vQ/EvQ/Ev - \ldots \\ &= v - vQ/Eg. \end{aligned} \tag{A6}$$

Here Q is a projection operator which prevents the particles in intermediate states being scattered into the Fermi sea since such terms would violate the Pauli exclusion principle and E is the usual perturbation theory energy denominator.

In lowest order the Brueckner, Bethe, Goldstone (BBG) theory yields for the ground state energy

$$\langle H \rangle = \sum_i \left(\phi_i \middle| -\frac{\hbar^2}{2m}\nabla^2 \middle| \phi_i \right) + \tfrac{1}{2}\sum_{ij} (\phi_i\phi_j|g|\phi_i\phi_j)_{AS}, \tag{A7}$$

i.e. the Hartree–Fock expression (A4) with the bare interaction v replaced by an effective interaction g. The formal expression

(A6) now becomes in detail

$$(ij|g|ij)_{AS} = (ij|v|ij)_{AS} - \sum_{ab} \frac{(ij|v|ab)(ab|g|ij)}{\varepsilon_a+\varepsilon_b-\varepsilon_i-\varepsilon_j} \quad \text{(A8)}$$

for the antisymmetrised matrix element in (A7), and the sum on a and b in (A8) is over single particle states not occupied in the ground state unperturbed Slater determinant. The occupied single particle states ϕ_i and energies ε_i are given by

$$-\frac{\hbar^2}{2m}\nabla^2\phi_i(\mathbf{r}) + \sum_j \int \phi_j^*(\mathbf{r}')g(\mathbf{r},\mathbf{r}')\Big(\phi_i(\mathbf{r})\phi_j(\mathbf{r}') - \phi_i(\mathbf{r}')\phi_j(\mathbf{r})\Big)d\mathbf{r}' = \varepsilon_i\phi_i(\mathbf{r}), \quad \text{(A9)}$$

i.e. again the Hartree–Fock-like equations (A3) with v replaced by g. The unoccupied state energies ε_a and ε_b are, however, not well defined and considerable controversy surrounds their definition – 'the dispersion correction'. A popular choice is to represent them by their kinetic energies and the sum on intermediate states then becomes a double momentum integral. The adequacy of this approximation has however been challenged by Pandharipande, V., Wiringer, R.B., and Day, B.D. (1976). *Phys. Lett.* **B 57**, 105; and the full BBG theory is reviewed in the three articles by Day, B.D., Bethe, H.A., Rajaraman, R., and Brandow, B. (1967). *Rev. Mod. Phys.* **39**.

In Hartree–Fock type calculations there is a particular attraction in using zero-range forces because of the simplicity that they introduce. In recent years a class of interactions known as Skyrme interactions (Beiner, M. *et al.*, *Nuc. Phys.* (1975). **A 238**, 29) has become popular. These interactions contain both two- and three-body terms

$$V_{\text{Skyrme}} = A\delta(\mathbf{r}_1-\mathbf{r}_2) + B\delta(\mathbf{r}_1-\mathbf{r}_2)\delta(\mathbf{r}_2-\mathbf{r}_3) \quad \text{(A10)}$$

where A and B may contain spin and/or isospin exchange operators. The Hartree–Fock mean field produced by such an interaction is given by

$$\langle i|V^{\mathrm{HF}}_{\mathrm{Skyrme}}(\mathbf{r}_1)|i\rangle = \sum_j \langle ij|A\delta(\mathbf{r}_1-\mathbf{r}_2)|ij\rangle_{\mathrm{AS}}$$

$$+ \sum_{j>k} \langle ijk|B\delta(\mathbf{r}_1-\mathbf{r}_2)\delta(\mathbf{r}_2-\mathbf{r}_3)|ijk\rangle_{\mathrm{AS}} \qquad \text{(A11)}$$

so that

$$V^{\mathrm{HF}}_{\mathrm{Skyrme}}(\mathbf{r}_1) = a\rho(\mathbf{r}_1) + b\rho^2(\mathbf{r}_1) \qquad \text{(A12)}$$

where the coefficients a and b depend on the specific exchange operators in A and B. Thus the mean field is a simple functional of the density.

Because of the similarity of lowest order BBG theory and Hartree–Fock theory, they are sometimes confused in the literature where terms like 'Brueckner–Hartree–Fock theory' abound. It should be stressed that Hartree–Fock theory rests on a variational principle and, as a consequence, the energies predicted by it are strictly upper bounds on the true energy of the system, while BBG theory is based upon perturbation theory and no variational principle is involved. The natural extension of Hartree–Fock theory to strongly interacting systems is the Jastrow approach described below.

A many-fermion trial variational wavefunction allowing two-body correlations can be constructed in the form:

$$\Psi = F\Phi \qquad \text{(A13)}$$

Again Φ is a Slater determinant (A2) and F is a symmetrised product of two-body correlation functions

$$F = \prod_{i>j} f_{ij} \qquad \text{(A14)}$$

The normalized expectation value of the hamiltonian may be written for an N-particle system

$$\frac{\langle\Psi|H|\Psi\rangle}{N\langle\Psi|\Psi\rangle} = \sum_{n=1}^{N} E_n \tag{A15}$$

where E_n is called the n-body cluster contribution to the energy and, in general, involves matrix elements of $3n$-fold spatial integrals for its evaluation. There are clearly many ways in which the cluster expansion can be ordered and summed. One particular scheme, known as the hypernetted chain method, is capable of summing very many orders in the cluster expansion at least for certain simplified forms of the correlation functions f. This is described in the papers of Pandharipande, V. and Bethe, H.A. (1973). *Phys. Rev.* **C 7**, 1312, and Fantoni, S. and Rosati S., (1975). *Nuovo Cim.* **A 25**, 593.

A useful low order approximation which can allow for the most general form of two-body correlation function has been developed by Owen, J.C., Bishop, R.F., and Irvine, J.M. (1976). *Ann. Phys. N.Y.* **102**, 170 and used widely in this work. This approach involves terminating the cluster expansion (A1) at the term E_2 and then carrying out a variational calculation to obtain the correlation functions f_{ij}. However, because the cluster expansion has been terminated, we no longer have a strict variational principle and hence a constraint $G([f])$ on the correlation functions is introduced which is intended to ensure the rapid convergence of the cluster expansion. Throughout this work, a constraint which ensures the normalization of the two-body wavefunctions has been employed

$$G([f]) = \rho\int d\mathbf{r}_{12}\left(1 - f^2(\mathbf{r}_{12})\phi^2(\mathbf{r}_{12})\right) = 1 \tag{A16}$$

where $\phi(\mathbf{r}_{12})$ is the Pauli correlated two-body relative wavefunction. Hence our constrained variational equations are

$$\frac{\delta}{\delta f}[E_2 - \lambda G] = 0 \tag{A17}$$

where λ is a Lagrange multiplier determined by the condition (A16). The total energy is then given by E_1, the usual Fermi gas energy

$$E_1 = \left(\frac{3}{10}\right)\frac{\hbar^2}{m}\,k_F^2 \tag{A18}$$

and

$$E_2 = \sum_{ij}\langle ij|\left(-\frac{\hbar^2}{2m}\Big[f(12),[\nabla_{12}^2,f(12)]\Big] + f(12)v(12)f(12)\right)|ij\rangle_{AS} \tag{A19}$$

In calculating the energy of nuclei in the crust region, the plane wave Slater determinant Φ is replaced by a shell model Slater determinant and for calculating the properties of a possible neutron-solid core, Φ is replaced by a Slater determinant of lattice functions.

Given the total energy $E(n)$ as a function of density n, the equation of state then follows from $P(n)$, the pressure:

$$P(\rho) = -n^2\,\frac{dE}{dn} \tag{A20}$$

APPENDIX B

PAIRING CORRELATIONS AND SUPERFLUIDITY

IN Appendix A we discussed methods of calculating the gross energy as a function of density for dense Fermi systems. From this the equation of state can be determined. While this is probably adequate for calculating the stability limits of neutron star masses and radii, it of course makes no allowance for more subtle phase changes which, while not making significant changes in the relationship between pressure and density, are extremely important for the structural or transport properties of neutron star material. So far we have only considered various fluid-to-solid transitions; we now consider pairing correlations within the fluid which can lead to a superfluid phase of the neutron star matter.

In Appendix A we did discuss the role of two-body correlations. However these were primarily of short range arising from the high momentum components of the interaction and yielded no long range co-operative phenomena (with the possible exception of the transition to the solid state). We now consider very long range correlations generated by extremely low momentum components of the interaction and specifically the zero momentum component. In a system with translational invariance the zero momentum two-body states are composed of pairs of particles in time-reversed motion where the particle with momentum **k** is coupled to the particle of momentum -**k**. If we consider as a model hamiltonian one in which the residual interaction, i.e. the interaction left over after the short range correlations of Appendix A have all been taken into account, only couples such paired states then it may be written in the form

$$H = \sum_{\mathbf{k}} \varepsilon_{\mathbf{k}} a_{\mathbf{k}}^{\dagger} a_{\mathbf{k}} + \sum_{\mathbf{kk}'>0} G_{\mathbf{kk}'} a_{\mathbf{k}}^{\dagger} a_{\mathbf{k}}^{\dagger} a_{-\mathbf{k}'} a_{\mathbf{k}'} \tag{B1}$$

where $\varepsilon_{\mathbf{k}}$ are the single particle states of Appendix A, $G_{\mathbf{kk}'}$ are the matrix elements of the residual interaction, while $a_{\mathbf{k}}^{\dagger}(a_{\mathbf{k}})$ is the occupation number representation creation (annihilation) operator for a particle in the state **k**. If the matrix elements $G_{kk'}$ are negative, i.e. the pairing interaction is attractive, then the ground state of the system will be composed of configurations in which all the particles are paired. A trial wavefunction for use in a variational calculation and composed entirely of such pairs was proposed by Bardeen, Cooper, and Schrieffer (BCS)

$$\Psi_{BCS} = \prod_{\mathbf{k}>0} (u_{\mathbf{k}} + v_{\mathbf{k}} a_{\mathbf{k}}^{\dagger} a_{-\mathbf{k}}^{\dagger}) |0\rangle \tag{B2}$$

where in order that Ψ_{BCS} be normalized we must have

$$u_{\mathbf{k}}^2 + v_{\mathbf{k}}^2 = 1 \tag{B3}$$

and $|0\rangle$ is the particle vacuum. Now Ψ_{BCS} is not an eigenstate of the number of particles and hence the variation on $v_{\mathbf{k}}$ must be carried out subject to the constraint

$$\langle \Psi_{BCS} | \sum_{\mathbf{k}} a_{\mathbf{k}}^{\dagger} a_{\mathbf{k}} | \Psi_{BCS} \rangle \doteq N \tag{B4}$$

i.e. the grand canonical ensemble approximation. Assuming that the $G_{\mathbf{kk}'}$ can all be approximated by an average matrix element $-G$ the variational equations

$$\frac{\delta}{\delta v_{\mathbf{k}}} \langle \Psi_{BCS} | H - \lambda \sum_{\mathbf{k}} a_{\mathbf{k}}^{\dagger} a_{\mathbf{k}} | \Psi_{BCS} \rangle = 0 \tag{B5}$$

yield

$$v_{\mathbf{k}}^2 = \tfrac{1}{2}[1 - (\varepsilon_{\mathbf{k}} - \lambda)\{(\varepsilon_{\mathbf{k}} - \lambda)^2 + \Delta^2\}^{-\frac{1}{2}}] \tag{B6}$$

where λ is the Lagrange multiplier associated with the constraint (B4), i.e. the chemical potential or Fermi energy. The pairing gap Δ is given by

$$\Delta = G \sum_{\mathbf{k}} u_{\mathbf{k}} v_{\mathbf{k}} \tag{B7}$$

and we note that

$$v_{\mathbf{k}}^2 = \langle \Psi_{BCS} | a_{\mathbf{k}}^{\dagger} a_{\mathbf{k}} | \Psi_{BCS} \rangle \tag{B8}$$

whence the identification of $v_{\mathbf{k}}^2$ as the probability that the single particle state $\phi_{\mathbf{k}}$ is populated in the BCS ground state. In Fig. (B1) we schematically represent the solution of (B6).

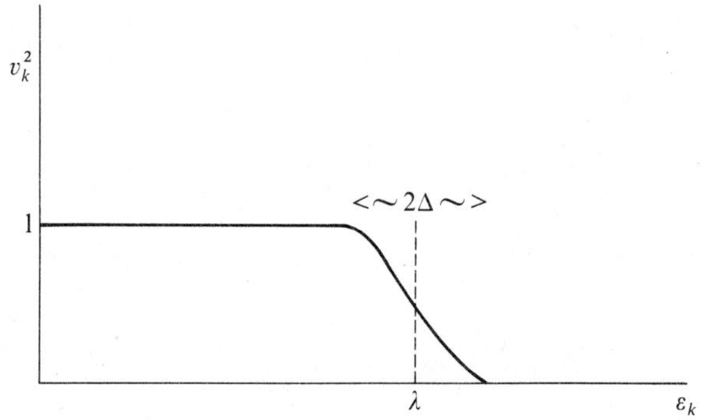

Fig. B1. The occupation probability v^2 of the state ϕ_k plotted against the single particle energy ε_k.

Inserting eqns. (B3) and (B6) into eqn. (B7) we obtain a dispersion relation for the pairing energy gap

$$\Delta = \tfrac{1}{2}\Delta G\sum_{\mathbf{k}}\{(\varepsilon_{\mathbf{k}}-\lambda)^2 + \Delta^2\}^{-\frac{1}{2}} \tag{B9}$$

which has as solutions, either $\Delta = 0$, i.e. the uncorrelated degenerate Fermi sea, or

$$\frac{2}{G} = \sum_{\mathbf{k}}\{(\varepsilon_{\mathbf{k}}-\lambda)^2 + \Delta^2\}^{-\frac{1}{2}}. \tag{B10}$$

Assuming a constant density of single particle states $n(\varepsilon)$ and converting the summation in (B10) into an integral we find, that for a weakly interacting system, i.e. $Gn \ll 1$, we have

$$\Delta = 2\omega\,\exp(-1/\eta G) \tag{B11}$$

where pairing correlations extend over a range $\lambda - \omega \leqslant \varepsilon \leqslant \lambda + \omega$. The effect of the correlations is to lower the energy of the ground state by an amount

$$\Delta E = \sum_{\varepsilon_{\mathbf{k}}<\lambda} \varepsilon_{\mathbf{k}} - \sum_{\mathbf{k}} (\varepsilon_{\mathbf{k}} - \lambda) v_{\mathbf{k}}^{2} + G(\sum_{\mathbf{k}} v_{\mathbf{k}} u_{\mathbf{k}})^{2}$$

$$\simeq \eta \Delta_0^2. \tag{B12}$$

The density of states at the Fermi surface for a non-interacting Fermi gas is

$$\eta = \frac{\pi^2}{2\hbar^2}\left(\frac{3n}{\pi^2}\right)^{1/3} \tag{B13}$$

where n is the particle number density.

Typically in the neutron fluid region the pairing energy ΔE of (B12) is ~ 1 per cent of the total interaction energy and hence makes little difference to the pressure–density relation. Where does the new physics come from which differentiates the normal from the BCS state? The BCS state is more stable against intrinsic excitations than the normal state. To see this we note that Ψ_{BCS} is the vacuum for the fermion quasiparticle operators

$$\alpha_{\mathbf{k}}^{\dagger} = v_{\mathbf{k}} a_{\mathbf{k}}^{\dagger} + u_{\mathbf{k}} a_{\mathbf{k}} \tag{B14}$$

and that in terms of these quasiparticles the hamiltonian (B1) is almost diagonal

$$H = \sum_{\mathbf{k}} E_{\mathbf{k}} \alpha_{\mathbf{k}}^{\dagger} \alpha_{\mathbf{k}} + \text{(terms with vanishing ground state expection value)}. \tag{B15}$$

The quasiparticle energies $E_{\mathbf{k}}$ are

$$E_{\mathbf{k}} = \left\{(\varepsilon_{\mathbf{k}} - \lambda)^2 + \Delta_0^2\right\}^{\frac{1}{2}} \tag{B16}$$

and hence, since the first excited state involves breaking a pair of particles and hence creating two quasiparticles, the minimum first excitation of the BCS state lies at ~ $2\Delta_0$. Conservation of energy then implies that there can be no dissipation of energy via friction or viscous modes until the collective energy density is sufficient to create quasiparticle excitations, i.e. the system is superfluid or, if charged superconducting, below some critical velocity.

At a finite temperature T the pairing gap will be $\Delta(T)$ and the various summations over states will have to take into account the Fermi distribution function

$$n(\varepsilon) = [\exp\{(\varepsilon-\lambda)/kT\} + 1]^{-1} \tag{B17}$$

We then find that the gap vanishes at a critical temperature T_c given by

$$kT_c = 0.57\ \Delta_0 \tag{B18}$$

For $kT \ll \Delta$ we have

$$\Delta = \Delta_0\{1 - (2\pi kT/\Delta_0)^{\frac{1}{2}}\exp(-\Delta_0/kT)\} \tag{B19}$$

while at temperatures $T \simeq T_c$

$$\Delta = 3.06\ kT_c\surd(1 - T/T_c). \tag{B20}$$

If we consider the fluid density ρ to be composed of a superfluid component ρ_s and a normal component ρ_n

$$\rho = \rho_s + \rho_n \tag{B21}$$

then the normal component is given by the occupation of nonzero momentum states

$$\rho_n = -\frac{4\pi}{3h^3}\int p^4\ \frac{dn}{d\varepsilon}\ dp \tag{B22}$$

whence

$$\begin{aligned} \rho_n/\rho &= (2\pi\Delta_0/kT)^{\frac{1}{2}}\exp(-\Delta_0/kT) \text{ for } kT \ll \Delta \\ &= (2T/T_c - 1) \text{ for } T \simeq T_c \end{aligned} \tag{B23}$$

Studies of nuclear systems suggest $T_c \sim 10^9$ K (see Fig. 3.1).

APPENDIX C

SOME RESULTS OF GENERAL RELATIVITY

THE Newtonian gravitational potential ϕ arising from a mass density distribution ρ is given by Poisson's equation

$$\nabla^2\phi = 4\pi G\rho \qquad \text{(C1)}$$

where G is the Newtonian gravitational constant and has the value 6.67×10^{-8} cm^3 g^{-1} s^{-2}. In the weak field limit the time–time component of the metric tensor is

$$g_{00} \simeq +\left(1 + \frac{2\phi}{c^2}\right) \qquad \text{(C2)}$$

while the corresponding component of the energy–momentum tensor is simply

$$T_{00} \simeq \rho c^2 \qquad \text{(C3)}$$

and thus eqn. (C1) could be rewritten

$$\nabla^2 g_{00} = \frac{8\pi G}{c^4} T_{00} \qquad \text{(C4)}$$

The simplest generalization of this for weak fields is

$$\mathbf{G}^{\omega}_{\mu\nu} = -\frac{8\pi G}{c^4}\mathbf{T}^{\omega}_{\mu\nu} \qquad \text{(C5)}$$

where $\mathbf{G}^{\omega}_{\mu\nu}$ is a linear combination of the metric and its first and second derivatives. The principle of equivalence then tells us that the gravitational field equations for a field of arbitrary strength must take the Einstein form

$$\mathbf{G}_{\mu\nu} = -\frac{8\pi G}{c^4}\mathbf{T}_{\mu\nu} \qquad \text{(C6)}$$

where $\mathbf{G}^{\omega}_{\mu\nu}$ and $\mathbf{T}^{\omega}_{\mu\nu}$ are the weak field limits of $\mathbf{G}_{\mu\nu}$ and $\mathbf{T}_{\mu\nu}$ respectively.

The $\mathbf{G}_{\mu\nu}$ can be determined from the requirements that (a) it is a tensor, (b) it only contains terms quadratic in the first derivatives and linear in the second derivatives of the metric, (c) since $\mathbf{T}_{\mu\nu}$ is symmetric $\mathbf{G}_{\mu\nu}$ must be symmetric, and (d) since $\mathbf{T}_{\mu\nu}$ is conserved so is $\mathbf{G}_{\mu\nu}$, i.e.

$$\mathbf{G}^{\mu}_{\nu,\mu} = 0 \text{ (covariant derivative zero).} \tag{C7}$$

Then[1]

$$\mathbf{G}_{\mu\nu} = \mathbf{R}_{\mu\nu} - \frac{1}{2}\mathbf{g}_{\mu\nu}R \tag{C8}$$

where $\mathbf{R}_{\mu\nu}$, the Ricci tensor, and R, the curvature scalar,

$$\mathbf{R}_{\mu\nu} \equiv \mathbf{R}^{\lambda}_{\mu\lambda\nu}, \quad R \equiv \mathbf{R}^{\mu}_{\mu} \tag{C9}$$

are contractions of the Riemann–Christoffel curvature tensor

$$\mathbf{R}^{\lambda}_{\mu\sigma\nu} \equiv \frac{\partial\Gamma^{\lambda}}{\partial x^{\nu}}\mu\sigma - \frac{\partial\Gamma^{\lambda}}{\partial x^{\sigma}}\mu\nu + \Gamma^{\eta}_{\mu\sigma}\Gamma^{\lambda}_{\nu\eta} - \Gamma^{\eta}_{\mu\nu}\Gamma^{\lambda}_{\sigma\eta} \tag{C10}$$

and the affine connection $\Gamma^{\lambda}_{\mu\nu}$ is defined in terms of the metric tensor $\mathbf{g}_{\mu\nu}$

$$\Gamma^{\lambda}_{\mu\nu} = \frac{1}{2}\mathbf{g}^{\eta\lambda}\left\{\frac{\partial\mathbf{g}_{\nu\eta}}{\partial x^{\mu}} + \frac{\partial\mathbf{g}_{\mu\eta}}{\partial x^{\nu}} - \frac{\partial\mathbf{g}_{\nu\mu}}{\partial x^{\eta}}\right\} \tag{C11}$$

From eqns. (C6) and (C7) we see that the field equations for empty space are

$$\mathbf{R}_{\mu\nu} = 0. \tag{C12}$$

For a static metric describing a spherically symmetric space, the solutions of (C6) can be shown to have the form

---

[1]We have here assumed that Einstein's cosmological constant is zero; see the books by Weinberg or Misner, Thorne, and Wheeler for details.

$$ds^2 = A(r)dt^2 - B(r)dr^2 - r^2 d\theta^2 - r^2 \sin^2\theta d\phi^2. \tag{C13}$$

The requirement that as $r \to \infty$ this metric should approach the Minskowski tensor in spherical polar co-ordinates imposes the condition

$$A(r) = c^2/B(r), \quad r \to \infty. \tag{C14}$$

For the empty space field equations (C12) there is a unique solution

$$A(r) = c^2[1 - 2MG/c^2 r] \tag{C15}$$

and, with this form, (C13) is known as the Schwarzschild metric and is the unique external solution to the Einstein equations for a static spherical mass distribution. We note that the Schwarzschild metric has a co-ordinate singularity at the Schwarzschild radius or event horizon

$$\mathcal{R} \equiv \frac{2MG}{c^2}. \tag{C16}$$

Another exact solution to equations (C12) is provided by the Kerr metric which looks like the external solution for a rotating mass

$$\begin{aligned} ds^2 &= c^2(1 - \mathcal{R}r/\rho^2)dt^2 + (2c\mathcal{R}ra \sin^2\theta/\rho^2)d\phi dt \\ &- (1 - \mathcal{R}r/\rho^2 + a^2\sin^2\theta/\rho^2)^{-1}dr^2 - \rho^2 d\theta^2 \\ &- (r^2 + a^2 + \mathcal{R}ra^2\sin^2\theta/\rho^2)\sin^2\theta d\phi^2 \end{aligned} \tag{C17}$$

where

$$\rho^2 \equiv r^2 + a^2\cos^2\theta \tag{C18}$$

and a is proportional to the angular momentum L

$$a \equiv L/Mc \tag{C19}$$

Unfortunately, the Kerr metric cannot be fitted smoothly to the internal solution for an arbitrary rotating star and thus may only be the solution for a rotating spherical black hole. Clearly, in the limit $L \to 0$ the Kerr metric tends to the Schwarzschild metric.

Given the metric the free fall orbit of a particle is given by the geodesic

$$\frac{d^2x^\mu}{ds^2} + \Gamma^\mu_{\nu\lambda}\,\frac{dx^\nu}{ds}\cdot\frac{dx^\lambda}{ds} = 0. \qquad \text{(C20)}$$

For the Schwarzschild metric there are two constants of the motion; first the 'angular momentum' per unit mass

$$a = l/mc = r^2 d\phi/ds \qquad \text{(C21)}$$

and second the energy per unit mass $e = E/mc^2$

$$e = \tfrac{1}{2}(dr/ds)^2 + (a^2/2r^2 - \mathcal{R}a^2/3r^3) - \mathcal{R}/2r \qquad \text{(C22)}$$

whence eqn. (1.45).

In a perfect fluid the energy–momentum tensor is

$$\mathbf{T}_{\mu\nu} = P\mathbf{g}_{\mu\nu} + (P/c^2 + \rho)\mathbf{u}_\mu\mathbf{u}_\nu \qquad \text{(C23)}$$

where P is the pressure and ρ is the relativistic mass density as measured by an observer in a locally inertial frame moving with the fluid at the instant of measurement and the velocity four-vector $\mathbf{u}^\mu$ of the fluid is defined so that

$$\mathbf{g}^{\mu\nu}\mathbf{u}_\mu\mathbf{u}_\nu = -1 \qquad \text{(C24)}$$

For a fluid at rest and the metric (C13) we have

$$u_r = u_\theta = u_\phi = 0;\; u_t = -[A(r)]^{-\frac{1}{2}} \qquad \text{(C25)}$$

Assuming time independence and spherical symmetry we find that Einstain's eqns. (C6) reduce to three equations for R_{rr}, $R_{\theta\theta}$, and R_{tt} which can be solved together with the condition of hydrostatic equilibrium for the metric functions $A(r)$, $B(r)$. Thus we find that

$$B(r) = (1 - 2GM(r)/c^2r)^{-1} \tag{C26}$$

where $M(r)$ is the mass inside the radius r

(C27)

$$M(r) = \int_0^r 4\pi\rho(r')r'^2\mathrm{d}r', \tag{C27}$$

$$\frac{1}{A(r)}\frac{\mathrm{d}A(r)}{\mathrm{d}r} = -\frac{2\mathrm{d}P/\mathrm{d}r}{[\rho c^2+P]} \tag{C28}$$

and the condition of hydrostatic equilibrium becomes (see eqn. (1.41))

$$\frac{\mathrm{d}P}{\mathrm{d}r} = -\frac{G[M(r) + 4\pi P(r)r^3/c^2][\rho(r) + P(r)/c^2]}{r[r - 2GM(r)/c^2]}. \tag{C29}$$

We see from eqn. (C26) how this metric joins smoothly to the Schwarzschild metric at $r = R$ and hence that the Schwarzschild metric is the unique solution to the Einstein equations outside a spherical, static mass source.

The total baryon number N of a star is conserved and given by

$$N = \int_0^R n_{\mathrm{B}}(r)[1 - 2GM(r)/c^2r]^{-\frac{1}{2}}4\pi r^2\mathrm{d}r \tag{C30}$$

where $n_{\mathrm{B}}(r)$ is the baryon number density as a function of radius. N is a measure of the amount of matter in the star and is a less ambiguous measure than the mass which includes the gravitational energy.

For slowly rotating stars (all velocities much less than c) we can develop a perturbation solution based on the Schwarzschild metric. We assume that, under the influence of the rotation, the star distorts and the pressure, density, and baryon number density become respectively $P + \Delta P$, $\rho + \Delta\rho$ and $n_B + \Delta n_B$ so that the modified energy–momentum tensor (see eqn. (C23)) now becomes

$$\mathbf{T}_{\mu\nu} = \mathbf{g}_{\mu\nu}(P+\Delta P) + \{(P+\Delta P)/c^2 + (\rho+\Delta\rho)\}\mathbf{u}_\mu\mathbf{u}_\nu \tag{C31}$$

We make a multipole expansion of the distortion effects assuming axial symmetry

$$\left.\begin{aligned} \Delta P &= (\rho+P/c\ )(p_0+p_2P_2(\cos\theta)) \\ \Delta\rho &= \Delta P(\mathrm{d}\rho/\mathrm{d}P);\ \Delta n_B = \Delta P(\partial n_B/\partial P). \end{aligned}\right\} \tag{C32}$$

Similarly the perturbed metric is expanded through second order in the rotational velocity Ω

$$\begin{aligned} \mathrm{d}s^2 &= A(r)\{1 + 2(h_0+h_2P_2(\cos\theta))\}\mathrm{d}t^2 - \\ &- B(r)\{1 + 2(m_0+m_2P_2(\cos\theta))/(r-\mathcal{R})\}\mathrm{d}r^2 - \\ &- r^2\{1 + 2(v_2-h_2)P_2(\cos\theta)\}\{\mathrm{d}\theta^2 + \sin^2\theta(\mathrm{d}\phi-\omega\mathrm{d}t)^2\} + \\ &+ 0(\Omega^3) \end{aligned} \tag{C33}$$

Here $P_2(\cos\theta)$ is the second Legendre polynomial and ω is the angular velocity of the local inertial frame and is proportional to Ω. This 'dragging' of the local inertial frame is known as the Lense–Thirring effect. The functions p_0, h_0, m_0, p_2, h_2, m_2, and v_2 are all functions of r proportional to Ω^2 which are obtained from the field equations. We need only obtain these functions for a single value of Ω the results for other values can then be obtained by scaling. The maximum angular velocity for which this approach should be valid is given by

$$\Omega^2 = GM/R^3 = \tfrac{1}{2}\mathcal{R}c^2/R^3. \tag{C34}$$

Above this angular velocity there will be a rotational shedding of mass.

We now define $\bar{\omega}$ the angular velocity relative to the local inertial frame

$$\bar{\omega} \equiv \Omega - \omega \tag{C35}$$

in terms of which the fluid inside the star moves with the four-velocity (compare with eqn. (C25))

$$\left.\begin{aligned} u_{\mathrm{r}} &= u_\theta = 0;\ u_\phi = \Omega u_t \\ u_t &= -A(r)^{-\frac{1}{2}}\{1 + \tfrac{1}{2}r^2\bar{\omega}^2 \sin^2\theta/A - h_0 - h_2P_2(\cos\theta)\} \end{aligned}\right\} \tag{C36}$$

The stellar equilibrium condition (C29) is now supplemented by the centrifugal forces and yields a condition for $\bar{\omega}$ from the field equation for $R_{t\phi}$

$$\frac{\mathrm{d}}{\mathrm{d}r}\left(r^4 J(r)\frac{\mathrm{d}\bar{\omega}}{\mathrm{d}r}\right) + 4r^3\,\frac{\mathrm{d}J(r)}{\mathrm{d}r}\,\bar{\omega} = 0 \qquad r < R \tag{C37}$$

where

$$J(r) = [A(r)B(r)]^{-\frac{1}{2}} \tag{C38}$$

and eqn. (C37) is solved subject to the boundary conditions that, at $r = 0$, $\bar{\omega}$ is regular and $\mathrm{d}\bar{\omega}/\mathrm{d}r\big|_{r=0} = 0$. The value of $\bar{\omega}(r{=}0)$ is determined by the condition

$$\bar{\omega}(R) = \Omega - 2LG/R^3c^2 \tag{C39}$$

where L is the 'total angular momentum' of the star

$$L = \frac{1}{6}\,\frac{c^2}{G}R^4\left(\frac{\mathrm{d}\bar{\omega}}{\mathrm{d}r}\right)_{r=R} \tag{C40}$$

and outside the star we have

$$\bar{\omega}(r) = \Omega - 2LG/r^3c^2 \qquad \text{(C41)}$$

The 'moment of inertia' I for the star is then defined to be

$$I = L/\Omega \qquad \text{(C42)}$$

The spherical deformations are given by the mass and pressure perturbation factors m_0 and p_0 (for $r < R$)

$$\frac{\mathrm{d}m_0}{\mathrm{d}r} = \frac{G}{c^2}4\pi r^2 \frac{\mathrm{d}\rho}{\mathrm{d}P}(\rho + P/c^2)p_0 + \frac{1}{12}J^2(r)r^4\left(\frac{\mathrm{d}\bar{\omega}}{\mathrm{d}r}\right)^2 - \frac{1}{3}r^3 \frac{\mathrm{d}J}{\mathrm{d}r}(r)\bar{\omega}^2 \qquad \text{(C43)}$$

$$\frac{\mathrm{d}p_0}{\mathrm{d}r} = -\frac{m_0(1+8\pi r^2P)}{(r-\mathcal{R})^2} - \frac{G}{c^2}\frac{4\pi r^2(\rho+P/c^2)}{(r-\mathcal{R})}p_0 + \frac{c^2}{12}\frac{r^4J^2(r)}{(r-\mathcal{R})}\left(\frac{\mathrm{d}\bar{\omega}}{\mathrm{d}r}\right)^2 +$$

$$+ \frac{c^2}{3}\frac{\mathrm{d}}{\mathrm{d}r}\left(\frac{r^3J^2(r)\bar{\omega}^2}{r-\mathcal{R}}\right) \qquad \text{(C44)}$$

Equations (C43) and (C44) solved with the boundary conditions $m_0(r=0) = p_0(r=0) = 0$ corresponds to a star with the same central density as the non-rotating one, which is a star with a greater mass than the original. If we call the mass of the rotating star $M + \Delta M$ then we have

$$\Delta M = \frac{c^2}{G}m_0(R) + GL^2/R^3c^4. \qquad \text{(C45)}$$

Outside the star we have

$$m_0(r) = \frac{G}{c^2}\Delta M - \frac{G^2}{c^6}L^2/r^3. \qquad \text{(C46)}$$

The remaining spherical function is then given by

$$h_0(r) = -\frac{1}{c^2}p_0(r) + \frac{1}{3}r^2A(r)^{-1}\bar{\omega}^2 + h_0(0) \qquad r < R$$

$$= \frac{G\Delta M}{c^2(r-\mathcal{R})} + \frac{G^2L^2}{c^2r^3(r-\mathcal{R})} \qquad r \geqslant R \qquad \text{(C47)}$$

where $h_0(0)$ is chosen to make $h_0(r)$ continuous at $r = R$.

Turning next to the quadrupole terms we have inside the star that v_2 and h_2 are given by

$$\frac{dv_2}{dr} = -\frac{1}{A(r)}\frac{dA}{dr}h_2(r) + \left(\frac{1}{r} + \frac{1}{2}\frac{dA/dr}{A}\right)\left[-\frac{1}{3}r^3\frac{dJ^2(r)}{dr}\bar{\omega}^2 + \frac{1}{6}J^2(r)r^4\left(\frac{d\bar{\omega}}{dr}\right)^2\right] \tag{C48}$$

$$\frac{dh_2}{dr} = \left\{-\frac{1}{A}\frac{dA}{dr} + \frac{r}{(r-\mathcal{R})}\frac{G}{c^2}\frac{A}{dA/dr}\left[8\pi(\rho+P/c^2) - \frac{4M}{r^3}\right]\right\}h_2(r) -$$

$$-\frac{4v_2}{r(r-\mathcal{R})}\frac{A}{dA/dr} + \frac{1}{6}\left[\frac{r}{2A}\frac{dA}{dr} - \frac{1}{(r-\mathcal{R})}\frac{A}{dA/dr}\right]r^3J^2\left(\frac{d\bar{\omega}}{dr}\right)^2 -$$

$$-\frac{1}{3}\left[\frac{r}{2A}\frac{dA}{dr} + \frac{1}{(r-\mathcal{R})}\frac{A}{dA/dr}\right]r^2\frac{dJ^2}{dr}\bar{\omega}^2 \tag{C49}$$

Solved with the boundary conditions $h_2(r=0) = v_2(r=0) = 0$. Outside the star the solutions are

$$h_2(r) = \frac{G^2}{c^6}L^2\left(\frac{2}{\mathcal{R}r^3} + \frac{1}{r^4}\right) + KQ_2^2\left(\frac{2r}{\mathcal{R}} - 1\right) \tag{C50}$$

$$v_2(r) = -\frac{G^2}{c^6}\frac{L^2}{r^4} + \frac{2KMG}{[r(r-\mathcal{R})]^{\frac{1}{2}}c^2}Q_2^1\left(\frac{2r}{\mathcal{R}} - 1\right) \tag{C51}$$

where $Q_n^m(\cos\theta)$ are the associated Legendre polynomials of the second class and K is a constant chosen to make h_2 and v_2 continuous at $r = R$. Finally, the non-radial mass and pressure factors m_2 and p_2 are given by

$$m_2(r) = (r-\mathcal{R})\left\{-h_2(r) - \frac{1}{3}r^3(dJ^2(r)/dr)\bar{\omega}^2 + \frac{1}{6}r^4J^2(r)\left(\frac{d\bar{\omega}}{dr}\right)^2\right\} \tag{C52}$$

and

$$p_2(r) = -c^2h_2(r) - \frac{c^2}{3}r^2\bar{\omega}^2/A(r). \tag{C53}$$

We may interpret these results as follows: a constant density spherical surface is distorted by rotation into the spheroid

$$r \to r + \xi_0(r) + \xi_2(r)P_2(\cos\theta) \tag{C54}$$

where

$$\xi_0(r) = -p_0(\rho+P/c^2)/dP/dr \tag{C55}$$

and

$$\xi_2(r) = -p_2(\rho+P/c^2)/dP/dr \tag{C56}$$

The spheroid has a mean radius

$$\bar{r} = r + \xi_0(r) \tag{C57}$$

and an eccentricity

$$e = \left[-3(v_2-h_2+\xi_2/r)\right]^{\frac{1}{2}}. \tag{C58}$$

The quadrupole moment of the star can be obtained from the coefficient of the $r^{-3}P_2(\cos\theta)$ term in the long-range asymptotic gravitational field (the Newtonian correction)

$$Q = \frac{8}{5}\frac{G^2}{c^4}KM^3 + L^2/Mc^2 \tag{C59}$$

For more details readers are referred to Misner, C.W., Thorne, K.S., and Wheeler, J.A., *Gravitation*, W. H. Freeman and Co., San Francisco and Weinberg, S. (1972). *Gravitation and cosmology*, John Wiley, New York.

INDEX